A LATE CRETACEOUS DINOSAUR AND REPTILE ASSEMBLAGE FROM SOUTH CAROLINA, USA

David R. Schwimmer, Professor of Geology, Department of Earth and Space Sciences, Columbus State University, Columbus, Georgia. Research interests include Late Cretaceous vertebrate paleontology and paleoenvironments of the Southeastern United States.

Albert E. Sanders, Curator of Natural Sciences, The Charleston Museum, Charleston South Carolina, retired. Research interests include the fossil vertebrates of South Carolina and, particularly, the evolution of cetaceans during Paleogene time.

Bruce R. Erickson, Fitzpatrick Chair of Paleontology, Department of Paleontology, The Science Museum of Minnesota, St. Paul, Minnesota. Major research interests are anatomy and paleoecology of Cretaceous and early Tertiary reptiles.

Robert E. Weems, United States Geological Survey, retired. Current research interests include Mesozoic and Cenozoic fish and reptiles, and vertebrate biostratigraphy of the eastern United States.

A LATE CRETACEOUS DINOSAUR AND REPTILE ASSEMBLAGE FROM SOUTH CAROLINA, USA

David R. Schwimmer
Albert E. Sanders
Bruce R. Erickson
Robert E. Weems

American Philosophical Society Press
Philadelphia • 2015

Transactions of the
American Philosophical Society
Held at Philadelphia
For Promoting Useful Knowledge
Volume 105, Part 2

ISBN: 978-1-60618-052-5

US ISSN: 0065-9746

Library of Congress Cataloging-in-Publication Data

Schwimmer, David R.
  A Late Cretaceous dinosaur and reptile assemblage from South Carolina, USA / David R.
Schwimmer, Albert E. Sanders, Bruce R. Erickson, and Robert E. Weems.
    pages cm. — (Transactions of the American Philosophical Society, ISSN 0065-9746 ;
volume 105, part 2)
    Summary: Discovery of Late Cretaceous dinosaur and reptile remains from Campanian
and Maastrichtian deposits in South Carolina. — Supplied by publisher.
    Includes bibliographical references and index.
    ISBN 978-1-60618-052-5 (alk. paper)
    1. Dinosaurs—South Carolina. 2. Paleontology—South Carolina. 3. Fossils—South
Carolina. 4. Paleontology—Cretaceous. 5. Campanian-Maastrichtian boundary. I. Sanders,
Albert E. II. Erickson, Bruce R. III. Weems, R. E. (Robert E.) IV. Title.
    QE861.8.S64S39 2015
    567.909757—dc23
                        2015012214

CONTENTS

PREFACE

This volume describes a new assemblage of Late Cretaceous dinosaur and reptile remains from Campanian and Maastrichtian deposits of eastern South Carolina. Six of the fourteen localities include new occurrences of theropod and hadrosaur dinosaurs, substantially increasing the known localities in the eastern United States that have produced dinosaur remains. Dinosaur remains were associated with abundant bones and skull fragments of the marine pleurodire (side-necked) turtle *Bothremys*, as well as marine and freshwater cryptodire turtles, including species of *Adocus*, *Osteopygis*, and Trionychidae, and, among the chelonioids, *Toxochelys*, *Peritresius*, *Euclastes*, and a new species of the dermochelyid *Corsochelys* described herein. Other reptilian remains in these localities are three identifiable crocodilian genera (*Deinosuchus*, *Bottosaurus*, and *Borealosuchus*) and an indeterminate longirostrine, mosasaurs, plesiosaurs, and a teiid lizard. This assemblage overall consists of fragmented bones and isolated teeth, characteristic of lag deposits, containing admixed marine and nonmarine taxa, dominated by turtles and crocodilians. Taphonomic and paleoecological aspects of the fauna are discussed, including coprolites and bite marks on bones.

The important but seldom-mentioned role that South Carolina played in the early history of studies of North American Cretaceous deposits and faunas is outlined, and the stratigraphic settings of fourteen localities are analyzed. Of the six localities where dinosaur bones were found, two principal sites yielded the majority: one at Stokes Quarry, Darlington County, in the mid-Campanian Coachman Formation; the other at Kingstree, Williamsburg County, from the late Campanian Donoho Creek Formation.

ACKNOWLEDGMENTS

This volume would not exist were it not for the foresight and efforts of Derwin Hudson and Frank Morning, Jr., of Florence, South Carolina, and Ray Ogilvie, of Hartsville, South Carolina, three avocational collectors who recognized the potential scientific value of the fossil vertebrate material that they began collecting in Florence and Darlington counties two decades ago. Their persistence in searching out localities and specimens has added more information to the fossil record of their region of the state than has ever been recorded before, and we are sincerely grateful to them for sharing their knowledge of the area with us, for their unflagging interest and assistance in the field, and for their contributions of the vast majority of the specimens used in this study.

Charleston Museum volunteer Billy T. Palmer collected many of our specimens, and his activities at the Kingstree locality shed new light on the vertebrate material found there. We also thank Bruce C. Lampright and Robert C. Melchior (Bemidji State University), who assisted with the recovery of specimens from the Kingstree site in 1984 and 1985. We are especially grateful to Mr. L. H. Stokes, of the Stokes Sand and Gravel Company, for permitting Ogilvie, Morning, and others associated with this project to collect fossils at Stokes Quarry while it was under his ownership, and to the new owner, Mr. William E. Dauksch, who graciously permitted Ogilvie and Morning to continue collecting activities at the site. Thanks are also extended to Karen Knight, Vance McCollum, David Cicimurri, Mike Bruggerman, and Eric Ogilvie for their fieldwork at the Stokes Quarry site. This book would not have been completed without the help of James L. Knight, formerly of the South Carolina State Museum, who arranged the loan of important specimens in that collection.

We are especially grateful to Dr. C. W. Clendinin, Director of the South Carolina Geological Survey, for authorizing the drilling of auger holes by Survey personnel to determine the age of the Cretaceous sediments at two of our sites, and to Ralph H. Willoughby of the Survey for his logs of the two holes. Thanks are due to Ralph H. Willoughby and to Will Doar of the Survey for informative comments on the drilling results, and to the drill crew, Joe Koch, Ernest H. Howard, and Quinton Jones. We sincerely appreciate the cooperation of the property owners, Mr. L. H. Stokes, of the Stokes Sand and Gravel Company, for permission to put down auger holes at Stokes Quarry, and Mr. Richard Byrd, of Turbeville, South Carolina, for allowing us to drill on the property where, nearly 50 years ago, Mr. Byrd's father, John C. Byrd, discovered dinosaur bones that had not been reported in the literature until their inclusion in the present work.

For deriving critical age determinations from pollen samples from the Cretaceous units at the Stokes Quarry, and Quinby and Turbeville localities, we are grateful to Raymond A. Christopher (formerly, US Geological Survey [USGS] and Clemson University), whose analyses provided great insight into the paleoenvironment of the deposits at those sites. For their advice on the construction of the stratigraphic chart and for their most helpful reviews of the geological section of this volume we thank Ray Christopher, David C. Prowell (USGS, retired), and Jean M. Self-Trail (USGS), who also examined sediment samples from the Kingstree locality but found them barren of nannoplankton. We also thank Lucy Edwards (USGS) and Ralph H. Willoughby for useful discussions on various aspects of this study. For their most helpful reviews of various drafts of the manuscript, we sincerely thank Philip W. Currie, Eugene S. Gaffney, Barbara Grandstaff, J. Howard Hutchison, David C. Parris, and Dale A. Russell.

Special acknowledgment is also extended to Lois Erickson, who assisted with laboratory analysis and the typing of previous drafts of the manuscript; to Randa Sanders, for help with the sorting of specimens; to Julie Martinez, who prepared the scientific drawings; and to Tim Erickson for figure-imaging work.

— Acknowledgments —

We take particular pleasure in thanking Mary McDonald, Editor, American Philosophical Society, for her great patience and help during the preparation of this work.

1

INTRODUCTION AND HISTORY OF UPPER CRETACEOUS COLLECTIONS IN SOUTH CAROLINA

Many locations in the eastern United States have produced an abundance of Upper Cretaceous vertebrate fossils from marine or paralic sediments (Zangerl 1948, 1953; Baird and Case 1966; Baird and Horner 1979; Bryan et al. 1991; Schwimmer 2002; Schwimmer et al. 1993; Gallagher 1993), but only a few occurrences of fossil findings have been reported from South Carolina. Upper Cretaceous coastline sites reconstructed by Schwimmer (1997a), record approximately twenty occurrences of Late Cretaceous dinosaurs, including two hadrosaurid teeth (SCSM [South Carolina State Museum] 87.158.1 and 87.158.2) from Kingstree, Williamsburg County, South Carolina, where some of the earliest discoveries of dinosaurs in South Carolina have been made. Other South Carolina dinosaur specimens were subsequently reported by Weishampel and Young (1996).

The collections of dinosaur and reptile remains discovered in South Carolina that are described here have been amassed principally from spoil piles of Upper Cretaceous (Black Creek Group) sediments at Stokes Quarry (Site 1) in Darlington County, and from stream banks at Quinby (Site 2), Black Creek (Site 5), and Burches Ferry (Site 8) in Florence County. Ten other localities in South Carolina have also provided specimens (Figure 1.1). Included in these collections are 61 dinosaur bones and teeth and an exceptional suite of reptile fossils. With few exceptions, all of the specimens are fragmentary, as is characteristic of lag accumulations (that is, loose material accumulated in pockets below flowing marine or freshwater). Many bones show evidence of modification by weathering and transport: for example, vertebral centra are numerous but generally lack neural arches or processes; long bones often are without one or both ends; and osteoderms and turtle shell bones are invariably incomplete. A few bones have been modified by predators or scavengers that left feeding traces. Three plesiosaur vertebrae from the Donoho Creek Formation are the only specimens that were found in association. Most other specimens lack direct stratigraphic information but in most cases it has been possible to deduce the general stratigraphic context of primary deposition for the specimens reported here.

Upper Cretaceous deposits at fourteen sites in Darlington, Florence, Clarendon, Williamsburg, Lee, and Horry counties (Figure 1.1) yielded the 275 specimens reported in this volume, which range in age from the

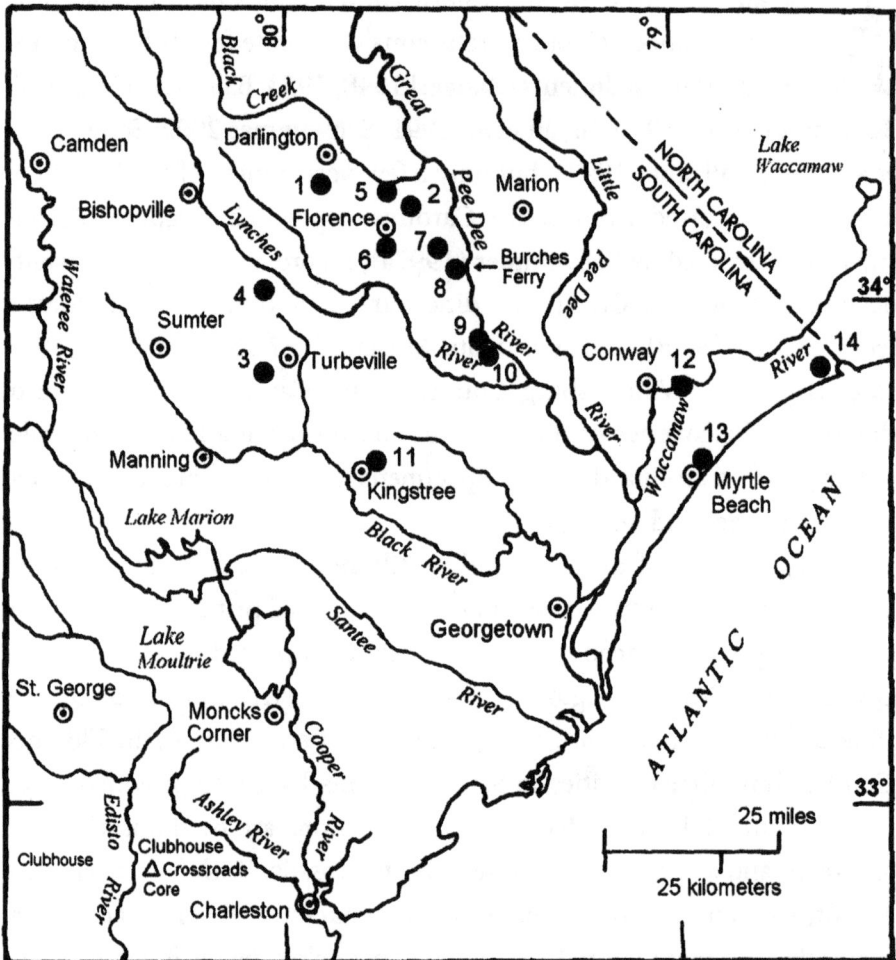

Figure 1.1 Localities in eastern South Carolina from which the Cretaceous reptile and dinosaur specimens reported here were found. Sites 1–2 are localities for which stratigraphic determinations were obtained through pollen analysis by R. Christopher in connection with this study. 1. Darlington Co., Stokes Quarry. 2. Florence Co., Quinby. 3. Clarendon Co., near Turbeville. 4. Lee Co., near Lynchburg. 5. Florence Co., Black Creek. 6. Florence Co., Florence, Muldrow's Mill. 7. Florence Co., Diamondhead Loop Road. 8. Florence Co., Burches Ferry. 9. Florence Co., Pee Dee River near Allison Ferry Landing. 10. Florence Co., Pee Dee River, two miles south of Allison Ferry Landing. 11. Williamsburg Co., Clapp Creek in Kingstree. 12. Horry Co., Waccamaw River. 13. Horry Co., Myrtle Beach. 14. Horry Co., near town of Little River.

mid-Campanian to late Maastrichtian stages of the Late Cretaceous. Marine taxa include chelonioid turtles (*Corsochelys*, *Toxochelys*, *Peritresius*, and *Euclastes*); pleurodire turtles; plesiosaurs; and mosasaurs, including tentatively identified genera *Tylosaurus* and *Prognathodon*. Predominantly freshwater turtle taxa are *Adocus*, *Osteopygis*, and two trionychid species. In addition, four crocodilians (*Borealosuchus*, *Bottosaurus*, *Deinosuchus*, and an indeterminate longirostrine) are present, representing habitats ranging from riverine to estuarine. *Deinosuchus* was a major near-shore predator that apparently preyed on hadrosaurs (Baird and Horner 1979; Schwimmer 2002; Rivera-Silva et al. 2009) as well as sea turtles (Schwimmer 1997b, 2002). Terrestrial taxa are represented by four dinosaur taxa (three theropods and a hadrosaurid), and a teiid lizard.

The geological and historical setting of the Coastal Plain where these sites are located is presented in Chapter 2. Collecting sites are documented with a brief history of each site and recognition of the numerous individuals responsible for collecting most of the specimens used in this study. Distribution of the taxa among the 14 sites is shown in Table 1.1.

Table 1.1   Distribution of Cretaceous Vertebrate Bones, Teeth, and Coprolites Found in the Fourteen Sites in South Carolina

Locality	Testudines	Plesiosauria	Teiid Lizards	Varanidae: Mosasaurs	Crocodylia	Dinosauria	Coprolites	Totals
Darlington Co., Stokes Quarry	66*	2	1	11	53*	24	7	164
Florence Co., Quinby	35	1		1	11			48
Florence Co., Burches Ferry	5	1		3	3	3		15
Williamsburg Co., Kingstree				1	4	10	1	16
Lee Co., quarry near Lynchburg	13							13
Clarendon Co., near Turbeville		1		1	2	1		5
Florence Co., Muldrow's Mill	2				1			3
Florence Co., 2 mi. south of Allison Ferry Landing		3 (assoc.)						3
Florence Co., near Allison Ferry Landing	1							1
Florence Co. Diamondhead Loop Road	1			1	1			3
Florence Co., Black Creek					1			1
Horry Co., Waccamaw River				1				1
Horry Co., Myrtle Beach				1				1
Horry Co., near Little River						1*		1
Totals	123	8	1	20	76	39	8	275

Asterisk (*) denotes inclusion of specimens with bite marks.

2

THE GEOLOGICAL AND HISTORICAL SETTING

The Coastal Plain of South Carolina is an eastward-thickening wedge along the continental margin of the state. The elevation of the top of the Cretaceous strata is 800 feet (243.8 m) below ground level, measured in the United States Geological Survey (USGS) Clubhouse Crossroads corehole, near Charleston (Gohn et al. 1977, 61, fig. 2; see Figure 1.1, p. 4). However, the Cretaceous sediments are only 5.3 feet (1.6 m) below ground at Burches Ferry on the Pee Dee River in Florence County (Self-Trail et al. 2002; see Figure 1.1). Consequently, the northeastern third of the Coastal Plain is the only region in the state in which exposures of Late Cretaceous sediments can be seen in the banks of certain creeks, rivers, and commercial quarrying operations.

Because of their accessibility, Cretaceous deposits of northeastern South Carolina have been a source of interest since the first quarter of the 19th century and were among the first sedimentary units of that age to be recognized along the Atlantic Coast of the United States. Not surprisingly, early studies of those deposits were hampered by misinterpretation and confusion with lithologically similar Tertiary units. Hence, an unwavering reliance on certain key index fossils, such as the oyster *Exogyra* and the fossilized rostral elements of belemnoid cephalopods, or "belemnites" (*Belemnitella*), was necessary to distinguish Cretaceous strata from Tertiary beds. Some early accounts are marred by incorrect locality data for Cretaceous specimens—they are recorded as having come from areas in which there are no exposures of Cretaceous rocks—those occasions often being the fault of collectors who did not keep adequate locality data with their specimens.

The first geological survey of South Carolina was conducted by Lardner Vanuxem in 1824, but limited funds restricted his work to five districts in the Piedmont region (Vanuxem 1826). Vanuxem made the first observations of the Cretaceous deposits of the state, and these were later included in a publication by Samuel George Morton (1829), apparently the earliest published reference to sediments of that age in South Carolina. Morton referred to such deposits as the "Secondary Formation," the term "Cretaceous" not being in general use at that time. He mentioned two localities in South Carolina that had yielded fossils of Cretaceous age, "Mars' Bluff, on Peedee river," and "Effingham's Mill, near the Eutaw Springs, on Santee river." According to Morton, belemnites were found at both localities, and "Exogyrae" at Effingham's Mill. Belemnites would be expected on the Pee Dee River, but there are no outcrops from the Cretaceous age on the Santee River, the calcareous units exposed there having been correctly recognized by Charles Lyell (1845) as being of Eocene age. Morton (1829, 63) stated that the purported discovery there, along with *Exogyra*, was "on the authority of Dr. Wm. Blanding." A physician-naturalist in Camden, South Carolina, William Blanding (1773–1857) corresponded with and sent specimens to Morton and other leading zoologists of the day (Sanders and Anderson 1999). Now a part of Orangeburg County, Eutaw Springs was in the old Charleston District in the 1820s, but the map of that district in *Mills' Atlas* (Mills 1825) does not show an "Effingham's Mill" near Eutaw Springs. As might be expected from the geology of the area, the location of "Effingham's Mill" near Eutaw Springs had been given in error. The true location of that mill—and the source of the fossils—became evident in one of Morton's most important publications.

Having gathered a considerable body of additional information about the Cretaceous deposits of the United States and the fossil remains that they had yielded, Morton published his landmark *Synopsis of the Organic Remains of the Cretaceous Group of the United States* in 1834. Containing 19 plates with 160 figures, it was the most comprehensive documentation of Cretaceous fossils from this country published to date. In that work Morton (1834) expanded his geographic coverage to include East Coast states from New Jersey and Delaware southward to Georgia and westward to Tennessee, Louisiana, Arkansas, and Missouri. His employment of the term "Cretaceous" is one of its earliest uses in the geological literature of North America. In this publication Morton (1834, 20) again includes "Mar's bluff, on Peedee river," and "Effingham's Mill" as localities in South Carolina that had furnished Cretaceous fossils. On this occasion, however, the location of Effingham's Mill is given as "Lynch's creek" instead of near Eutaw Springs, as in Morton's publication of 1829. On the map of the old Darlington District in *Mills' Atlas* (Mills 1825) the name "Effingham" is shown on the north side of "Lynches Creek" (now the Lynches River) next to the "Old Saw Mill Ferry" road, which is the present-day U.S. Route 52. This locality, now a part of Florence County, is affirmed by Tuomey (1848, 138) as the site of "Effingham's Mills, and was first made known by Dr. Blanding," the source of the fossils cited by Morton (1829,

1834). In his description of the tube shell of a new polychaete worm, *Hamulus onyx*, Morton (1834, 73, pl. 2, fig.8) reported and illustrated a tube shell of this form "obtained by Dr. Blanding at Lynch's Creek, South Carolina." The exact locality was not given, but in all probability Blanding collected the specimen from the Peedee Formation at the Effingham's Mill site on Lynches River, where, as noted by Morton (1834, 20), he had also collected *Exogyra costata*. Two specimens of *Hamulus* were collected from the Peedee at Burches Ferry by a Charleston Museum party in June 2006.

Morton (1834, 49, 72) also described and figured a new gastropod, *Conus gyratus*, and a new barnacle, *Balanus peregrinus*, collected from Cretaceous beds in South Carolina by Timothy Conrad, probably somewhere along the Pee Dee River. The holotypes of these two taxa are in the Academy of Natural Sciences of Philadelphia.

In addition to "Mars bluff" and "Effingham's Mill," Morton (1834) also reported Nelson's Ferry on the Santee River as a locality for Cretaceous fossils, "the *Belemnites Americanus*" (*Belemnitella americana* [Morton 1834]) supposedly having been collected there. Then, in the portion of the old Charleston District that is now a part of Orangeburg County, Nelson's Ferry was located a short distance north of Eutaw Springs (Mills 1825; Charleston District map), but that site now lies beneath the waters of Lake Marion. Nevertheless, in this region there are no outcrops of Cretaceous beds from which belemnites could have been collected. Once again, we see evidence of confusion of localities where specimens were collected. The belemnites purportedly collected at Nelson's Ferry probably came from a locality on the Pee Dee River, possibly Burches Ferry.

The earliest detailed descriptions of Upper Cretaceous sediments in South Carolina are those of Ruffin (1843, 24), who applied the name "Peedee bed" to outcrops of a marine formation on the Pee Dee River at "Britton's ferry," "Giles' bluff," "Birch's ferry," and other localities along the river. He recognized belemnites and the oyster *Exogyra costata* as characteristic fossils of that unit, noting that they were particularly abundant at "Birch's ferry." Ruffin (1843, 26–27) also observed that "By the presence of either of these well-marked fossils elsewhere . . . this particular Peedee bed may be readily and certainly determined." In a preceding paragraph he alluded to the distinctive lithology of the unit, stating that "this bed is so peculiar and uniform in appearance, that after it has been identified by the presence of the above fossils, it can be as well known even where they are altogether absent" (Ruffin 1843, 26).

In his *Report on the Geology of South Carolina*, Tuomey (1848) did not use Ruffin's name, "Peedee bed," but referred to those deposits as the "Cretaceous formation." Visiting the Cretaceous localities in South Carolina during his geologic survey of the state from 1845 to 1847, Tuomey (1848) gave descriptions of the lithology of those localities and noted the fossil taxa that he found at most of them. At "Birch's Ferry" on the Pee Dee River, he observed that "the soft, dark gray marl . . . for the first ten feet, is filled with *Belemnites*, which project from it in every direction. It looks as if some persons had amused themselves by driving as many as possible into the face of the bluff. Hundreds have fallen down and are strewed along the edge of the water" (Tuomey 1848,136). Sixteen specimens of *Belemnitella americana* cataloged as (ChM [The Charleston Museum] PI18339-18354) collected on the Pee Dee River by Francis S. Holmes are in the collection of The Charleston Museum. Holmes accompanied Tuomey on many of his field trips in connection with the geologic survey, and it is most probable that he collected those specimens at Burches Ferry on the occasion of Tuomey's (1848) observations of the abundance of *Belemnitella* at that locality.

As Tuomey (1848, 137) discovered, "It is quite evident that the Cretaceous formation underlies the whole country between the Little Peedee [River] and Lynch's Creek." He also noted that "Between Mar's Bluff and Darlington Court House, I observed it on Black Creek, where it is covered by sand, and at other points along the road. Near the village [Darlington] it underlies the Tertiary beds, and is exposed in thick fissile beds, at Col. Ervin's marl pit." Tuomey (1848, 138–39) provided the most comprehensive faunal list that had been reported from the Cretaceous strata of South Carolina to that date, including sixteen mollusks, two of which were cephalopods, and the first two vertebrate fossils from those strata, these being teeth of the sharks "*Carcharodon*, sp? near *megalodon*" and "*Lamna*, sp?" Of course, if the tooth "near *megalodon*" came from a Cretaceous unit, we can be sure that it was not referable to that large Miocene species and was not even assignable to *Carcharodon*. The whereabouts of these specimens today are unknown, so it is unlikely that we will ever know their true affinities. If they were in fact from Cretaceous sediments they would stand as the first published records of Cretaceous vertebrate fossils from South Carolina. However, the fact that one of them was compared to *Carcharodon* (= *Carcharocles*) *megalodon* is cause for some suspicion about their true stratigraphic origin.

Shortly after the publication of Tuomey's (1848) volume, the physician-naturalist Robert W. Gibbes, of Columbia, South Carolina, reported new records of mosasaur remains from the United States and described several new taxa (Gibbes 1850, 1851). Noting that mosasaur remains were first recorded in the United States by Samuel L. Mitchill (1818), Gibbes (1851) reviewed subsequent reports of these large marine lizards in the United States, none of which were from South Carolina. He then described three new species of *Mosasaurus*: *M. minor* from Alabama, *M. couperi* from Georgia, and *M. caroliniensis* from South Carolina, and three new "mosasauroid" genera, *Holcodus, Conosaurus,* and *Amphorosteus* (Gibbes, 1851).

Gibbes's (1851, 4, 7–8; pl. 2) *Mosasaurus caroliniensis* (see Figure 4.18 D, E, p. 79) was based on a fragment of a dentary (curiously called "lower maxilla" by Gibbes) found in the vicinity of Darlington, South Carolina, sent to him by Chancellor E. G. Dargan in 1848.

> It is reported as found with cetacean remains among the shells of the Pliocene. In Darlington, as the beds of the Pliocene rest upon the Cretaceous, it is most probable it was derived from the latter formation. Its appearance and the mineralization of its structure render it probable that it came originally from the Cretaceous. (Gibbes 1851, 7)

That specimen, a partial mosasaur dentary with a single tooth base in place (see Figure 4.20 C, p. 84), was noted by Gibbes (1851, 7) as being seven inches (17.8 cm) in length. Gibbes's (1851) conclusion that the specimen was of Cretaceous origin was correct.

His observation that in the Darlington area "the beds of the Pliocene rest upon the Cretaceous" also coincides with the stratigraphy at our Stokes Quarry locality in Darlington County (Site No. 1, see Figure 1.1, p. 4), where Pliocene sediments overlie the Cretaceous deposits. Gibbes's (1851) description of *Mosasaurus caroliniensis* is the first report of Cretaceous vertebrate material from the region covered in the present study and is also the first record of mosasaurs from South Carolina. Gibbes (1851, 8) remarked that "From the size of the tooth and estimated thickness of the maxilla [i.e., dentary], it must have characterized one of the largest species." Subsequently, Leidy (1865) synonymized *M. caroliniensis* with *M. mitchillii*, which in turn was referred to as *Tylosaurus* by Williston (1897). More recent allocations of this specimen have not been pursued because of the fragmentary nature of the specimen and because it has been irretrievably lost, as discussed in the text that follows.

Of the three genera of mosasaurs erected by Gibbes (1851)—*Holcodus, Conosaurus,* and *Amphorosteus*—only one, *Conosaurus,* was based on specimens from South Carolina. Gibbes (1851, pl. 3, figs. 1–9) noted six teeth, ranging from about 9 mm to 23 mm in crown height, to which he applied the name *Conosaurus bowmani,* noting that "I am indebted to Captain A. H. Bowman of the United States Topographical Engineers for several teeth of an acrodont saurian found in the Eocene of Ashley River, South Carolina" (Gibbes 1851, 9), a few miles north of Charleston. Immediately, there are major stratigraphic problems with Gibbes's consideration of those specimens as mosasaur teeth. The beds regarded as Eocene at that time—the present-day Ashley Formation—are now known to be of late Early Oligocene age (Weems et al. 2006), but in either case they are much too young to contain the teeth of mosasaurs. Furthermore, Cretaceous sediments are 745 feet (227 m) below ground level in the nearby city of Charleston (Bybell et al. 1998), placing mosasaur remains well out of reach. Not surprising, Leidy (1868) concluded that Gibbes's specimens were fish teeth and not mosasaur teeth and accordingly established the name *Conosaurops* to replace Gibbes's *Conosaurus.* Gibbes (1851) also mentioned "mosasaur" vertebrae from "the Eocene of Ashley River" (i.e., the Ashley Formation), but these undoubtedly were cetacean vertebrae, which are commonly found in those beds.

As inferred by Gibbes (1851), the specimens on which he founded his three new species of *Mosasaurus* and his three new "mosasaur" genera were in his extensive private collection of fossils. Unfortunately, that collection, along with his many other valuable possessions, was lost when Gibbes's home was consumed by flames during the burning of Columbia during the occupation of that city by Sherman's army in February 1865 (Sanders and Anderson 1999). Though none of the six names that he proposed are valid today, Gibbes's (1851) paper is a highly useful reference on studies of North American mosasaurs in the mid-19th century and is a particularly valuable chronicle of specimens being found in South Carolina at that time. It appears, however, to be the last work published on the Cretaceous Period of South Carolina in the 19th century. The Civil War and the hard times of Reconstruction and its aftermath brought scientific enterprises in South Carolina to a virtual standstill during the remainder of that century.

After Tuomey's (1848) volume, nothing further was published about Cretaceous deposits in South Carolina until Sloan (1907) defined and applied the name "Black Creek shale" to a unit of soft shale and black clay exposed along Black Creek in Darlington and Florence counties (see Figure 1.1, Site 5, p. 4). Almost simultaneously, Stephenson (1907) gave the name "Bladen formation" to exposures of the same beds in Bladen County, North Carolina. Sloan's name had priority, however, because, even though he did not formally define the name until 1907, he had used it previously in an identical context on a geologic map issued in 1906 (USGS Geologic Names Committee meeting minutes, 1909, 802). Sloan's Black Creek shale was clearly the black laminated clay and shale that Tuomey (1848, 136) observed at Mars Bluff on the Pee Dee River, where it was also recorded by Sloan (1908, 443).

Subsequently, Sloan (1908, 442) renamed his Black Creek shale as "Black Creek Phase" and designated the younger beds overlying them (Ruffin's [1843] "Peedee bed" on the Pee Dee River) as the "Burches Ferry Phase," thereby recognizing the presence of two distinct Cretaceous stratigraphic units in the Pee Dee region of South Carolina.

Stephenson (1912) employed the more conventional name "Black Creek formation" instead of Sloan's (1908) Black Creek Phase, and to Sloan's (1908) Burches Ferry Phase he applied the name "Peedee sand," a reflection of Ruffin's (1843) earlier name, "Peedee bed." Stephenson (1923) later renamed the Peedee sand the "Peedee formation" without explanation. Stephenson's (1907) earlier name, "Bladen," was reinstated by Heron (1958) as the "Bladen member" in the lower part of the Black Creek Formation. Owens (1989) later mapped the Black Creek as a group consisting of a lower Tar Heel Formation, a middle Bladen Formation, and an upper Donoho Creek Formation (Figure 2.1). That stratigraphy was formalized by Sohl and Owens (1991), who described two new formations, the Tarheel in North Carolina and the Donoho Creek in South Carolina, and restricted the Bladen Formation to the upper marine part of the Bladen Member of Heron (1958). Gohn (1992) expanded the Black

MA	EPOCH	STAGE	S.C. FORMATIONS	CC ZONES	POLLEN ZONES	STUDY SITES
65	LATE CRETACEOUS (part)	Maastrichtian	Sawdust Landing	26 b	Su	
				26 a		— Turbeville Site
			Steel Creek / Peedee	25	Hc / Ct	— Burches Ferry
70						
75		Campanian	Donoho Creek	22	B	— Burches Ferry / Quinby Site
			Bladen	21	Upper	
			Coachman	20	Middle (C)	— Stokes Quarry
			Cane Acre	19	Lower	
80			Caddin	18	D	
83.5			? Shepherd Grove			

(Black Creek Group; Tar Heel (N.C.))

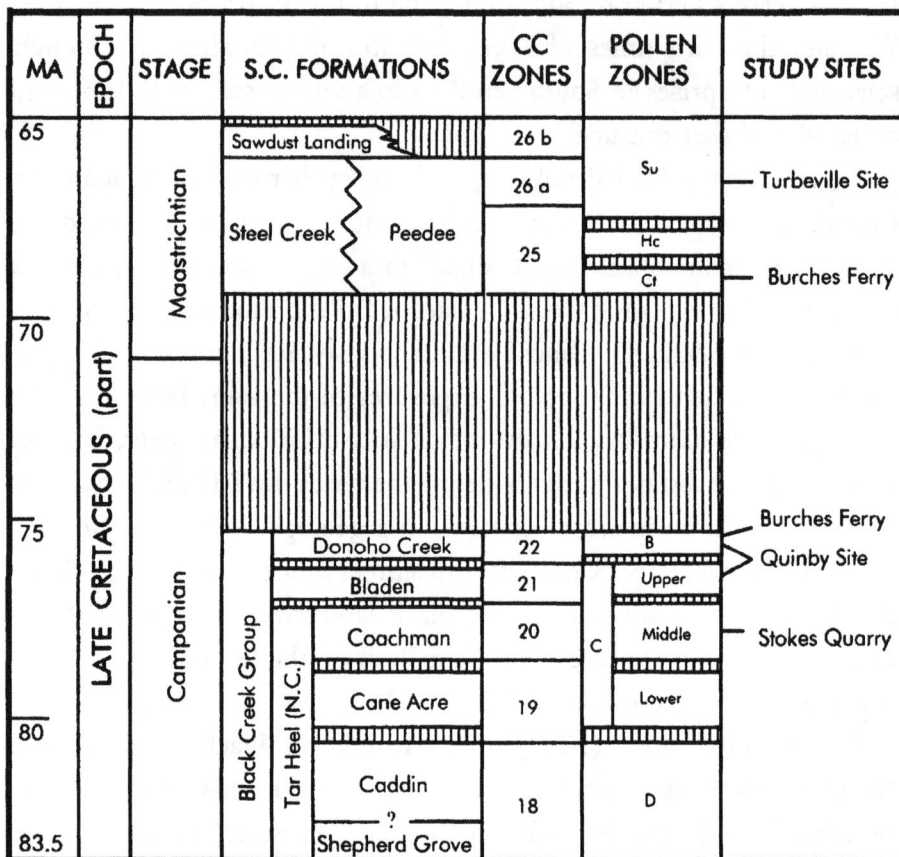

Figure 2.1 Chronostratigraphic sequence of Late Cretaceous deposits in South Carolina correlated with nannoplankton zones, pollen zones, and vertebrate fossil sites reported in this study. Pollen and nannofossil data from Sissingh (1977) and Christopher and Prowell (2002, 2010). Geologic age data from Gradstein et al. (2012). Stratigraphic positions of vertebrate fossil material at the Stokes Quarry, Quinby, and Turbeville sites were determined from pollen samples examined by R. A. Christopher and dated to horizons that are age-equivalents of units in sources above; however, the sampled strata are not referable to those units with certainty. Pollen date from the Quinby site was obtained from an outcrop on stream bank, but fossils came from a lag deposit at the base of the outcrop. Burches Ferry specimens were collected directly from the Donoho Creek and Peedee Formations, respectively. The Middendorf Formation is omitted from the chart because of its problematical stratigraphic position relative to other Late Cretaceous units in South Carolina (see Prowell et al. 2003, 63).

Creek Group to include the Cane Acre and the Coachman Formations, two stratigraphic units newly recognized in the Clubhouse Crossroads core in Dorchester County, South Carolina. Gohn (1992) also recognized and named two older subsurface units, the Shepherd Grove and Caddin Formations, in the Clubhouse Crossroads core. The six formations comprising the Black Creek Group (Shepherd Grove, Caddin, Cane Acre,

Coachman, Bladen, Donoho Creek [Prowell et al. 2003, fig.6]) and the Peedee, Steel Creek, Sawdust Landing, and Middendorf Formations are the Campanian and Maastrichtian stratigraphic units currently recognized in South Carolina. Regionally, those units are marine in the eastern part of South Carolina (Shepherd Grove, Caddin, Coachman, Bladen, Donoho Creek, Peedee) and are fluvial in the upper Coastal Plain and western part of the state (Steel Creek, Sawdust Landing, Middendorf; Prowell et al. 2003). The Tarheel Formation is generally regarded as the North Carolina equivalent of the Shepherd Grove, Caddin, Cane Acre, and Coachman Formations (J. Self-Trail, personal communication to Albert E. Sanders, 2006).

Recent collecting activities, especially in Florence and Darlington counties, have produced a remarkable body of vertebrate material ranging from the mid-Campanian to late Maastrichtian periods in age. Most of the material used in the present study was collected through intensive searches by avocational collectors, principally Derwin Hudson and Frank M. Morning, Jr., of Florence, and Ray Ogilvie of Hartsville, South Carolina. Working as the Midstate Geological Research Team, these three individuals systematically located many of the sites shown in Figure 1.1, plotted them on maps, collected specimens from them, and brought them to our attention. Billy T. Palmer, a volunteer with The Charleston Museum, also made significant recent additions to our study material. Hence, this book is largely the result of their efforts, as well as those of several other collectors mentioned in the following descriptions of the localities that yielded the material reported in this study.

3

COLLECTION SITES

T he present work reports Cretaceous dinosaur and reptile material from the following fourteen localities in Darlington, Clarendon, Florence, Horry, Lee, and Williamsburg counties. The numerical sequence of the sites (see Figure 1.1, p. 4) begins with the three sites (1, 2, 3) for which stratigraphic determinations were made through pollen analysis by Ray Christopher from sediment samples obtained in connection with the present study. Thereafter, the numbering roughly follows the geographical distribution of localities.

SITE NUMBER 1: STOKES QUARRY

S.C., Darlington County; south side of County Road 407, ca. 0.25 mile (0.40 km) west of County Road 49, 2.85 miles (4.58 km) east of. S.C. Route 340.

USGS Florence West 7.5' quadrangle, 34° 14' 21.8" N, 79° 30.0" W 50'.

Formation and Age

Coachman Formation; mid-Campanian, Late Cretaceous.

Determined from pollen extracted from augur-hole samples examined by R. Christopher and referred to middle zone "C" of his then-current pollen zonation scale (but see comments to follow about revised pollen zonation). From those samples Christopher made the following observations (R. Christopher, personal communication to Albert E. Sanders, 2005):

> Both samples are assigned to the informally proposed middle part of zone "C." This is one of the most difficult of the Upper Cretaceous zones to recognize because of the few marker, guide, and index fossils associated with it. However, the presence of the form referred to as CP3D-2 by Wolfe (1976) indicates assignment of the samples to middle zone "C" as this palynomorph is restricted in its stratigraphic range to this zone.
>
> Dinoflagellate cysts are very rare in both samples, and terrestrial organic matter dominates the residue. For this reason, it appears that the samples were deposited in a nonmarine environment, but with some marine influence. Elsewhere on the Coastal Plain, middle zone "C" has been observed in the Coachman Formation.

The palynological biozonation has been subsequently updated for South Carolina (Christopher and Prowell 2010), which would presumably place the same horizons in the Coachman Formation in the current *Holkopollenites forix* (Hf) Palynological Zone (J. Self-Trail, personal communication to David R. Schwimmer, 2014).

Description and History

One of the most important sources of Cretaceous vertebrate fossils on the East Coast of the United States, this locality was a small sand quarry operated by the Stokes Sand and Gravel Company until the property was sold to William E. Danksch in early 2006. The area that produced the vertebrate fossils was in a small space adjacent to the east end of the quarry, where specimens were found by surface collecting among piles of Cretaceous spoil piles left behind after commercial quarrying operations along the north bank of a large flooded pit created from earlier excavations at the quarry (Figure 3.1A). The Cretaceous sediments are overlain by Pliocene marine deposits containing a rich assemblage of marine mollusk shells, foraminifera, and ostracods, some of which indicate a pre-Duplin (late early Pliocene to early middle Pliocene) age for this unit (L. D. Campbell, personal communication to Albert E. Sanders, 2006). To obtain samples of the Cretaceous deposits for the present study, Ralph Willoughby, Will Doar, and others of the South Carolina Geological Survey, accompanied by the fourth author, put down an auger hole on the floor of the quarry approximately 10 ft (3.05 m) below the original ground surface and 5 or 6 ft (1.5–1.8 m) above the water level of the pit. The drill log recorded 14 additional feet (4.3 m) of Pliocene sediments overlying 16 ft (4.88 m) of Upper Cretaceous sediments (Figure 3.1B). The latter deposits are described in the drill log as: "Stiff, black . . . silty clay matrix with common to abundant dark greenish grey (5GY 4/1) well sorted, very fine to fine quartz sand in flaser beds a few mm to 2 cm thick. The very fine-to-fine sand dominates locally. Also with minor very fine glauconite in the sand laminae to thin beds" (see Figure 3.1B).

As noted previously, the source of the many Cretaceous vertebrate elements extracted from this locality was restricted to a relatively small area along the north bank of the pit near its east end. In this area

Figure 3.1A Spoil piles at the northeast end of a water-filled excavation at Stokes Quarry, ca. 5.6 kilometers south-southeast of Darlington, Darlington County, South Carolina (Site No. 1, Figure 1.1). Gray sediments yielded the mid-Campanian (Late Cretaceous) vertebrate remains reported here. Orange sediments are Pliocene sediments of pre-Duplin age. Among the more productive Late Cretaceous vertebrate sites on the East Coast of North America, this locality furnished 146 of the 275 specimens included in this study.

approximately 5 feet (1.5 m) of a local channel at the base of the Pliocene unit contained very abundant black phosphate clasts and bone material. Examination of the sediments in the pit left little doubt that the bulk of the black phosphatic material is derived from the reworked, concentrated lag in the base of the Pliocene deposits. In his remarks accompanying the drill log, R. Willoughby noted also that, "These materials were concentrated from the underlying Upper Cretaceous sediments during initial transgressions of the Pliocene . . . sea or perhaps, in part, in fluvial channels immediately prior to [Pliocene] marine deposition" (R. Willoughby, personal communication to Albert E. Sanders, 2005). As constituents of the lag material, the bones of Cretaceous age are assumed to have undergone transport and abrasion in this channel before final burial. Some

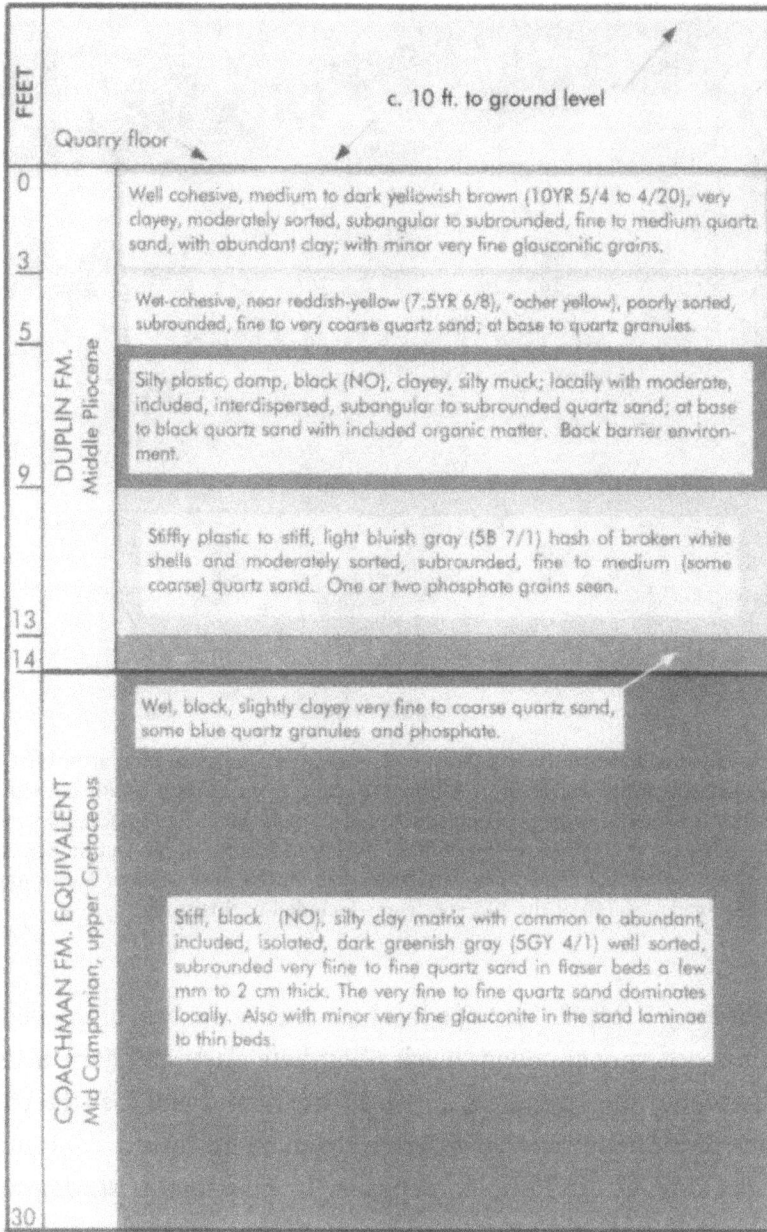

Figure 3.1B A stratigraphic section of mid-Campanian (Upper Cretaceous) Coachman Formation equivalent sediments unconformably overlain by marine deposits of the late Pliocene Duplin Formation in Stokes Quarry, south side of County Road 47, Darlington County, South Carolina (Site No. 1, Figure 1.1). Section reconstructed from samples from South Carolina Geological Survey power augur hole put down in quarry floor. Specimens reported were found in spoil piles of sediments dug by machinery from the upper portion of Campanian deposits.

specimens show considerable wear on broken surfaces, indicating transportation over some distance, whereas other less worn bones evidently had more local origins.

The presence of Cretaceous fossils at Stokes Quarry was first discovered by Frank M. Morning, Jr. and Ray Ogilvie, and the vast majority of specimens from this site were collected by them over a period of about four years. Additional specimens were collected by Mike Bruggerman, James L. Knight (all in the SCSM [South Carolina State Museum] collection), Eric Ogilvie, and Bill Palmer (ChM [The Charleston Museum] collections).

Stokes Quarry is particularly remarkable for the number of dinosaur specimens that it has produced, one of the largest bodies of such material known from a single locality on the East Coast of the United States. This site has also furnished two multituberculate specimens now under study for a separate publication, as well as a considerable body of fish remains.

SITE NUMBER 2: QUINBY

S.C., Florence County; small tributary of Black Creek, ca. 0.10 mile (0.16 km) east of County Road 1510 (Sand Pit Road), 1.3 miles north of County Road 358 near Quinby.

USGS Florence East 7.5′ quadrangle, 34° 14′ 44″ N, 79° 43′ 00″ W.

Formation and Age

Bladen Formation; mid-Campanian, Late Cretaceous.

A sediment sample taken by R. Ogilvie from the Cretaceous outcrop in the creek bank where specimens were collected yielded pollen referred by R. Christopher to Zone "B" of his pollen zonation scale (R. Christopher, personal communication to Albert E. Sanders, 2005). Christopher gave the following summary:

> The sample contains several species that range no higher than the top of the Campanian: *Osculapollis aequalis, Holkopollenites irregularis, Plicapollis usitatus*. Although none of these species is restricted to the late Campanian, no taxa were observed that are restricted to older parts of the section.

Figure 3.2 Outcrop of the laminated shaly clays of the late Campanian Donoho Creek Formation along a small tributary of Black Creek, ca. 1.6 kilometers, northwest of Quinby in Florence County, South Carolina (Site No. 2, Figure 1.1). This locality furnished Cretaceous vertebrate specimens reworked from the underlying Bladen Formation.

> Therefore, the sample is assigned to zone "B." The organic residue is dominated by terrestrial elements, although rare dinoflagellate cysts are present, indicating deposition in a nonmarine environment with some minor marine influence.

In later discussions with the second author, Christopher noted that Zone "B' is known to occur in the Donoho Creek Formation. That information, coupled with David Prowell's (United States Geological Survey [USGS], retired) observations of the Donoho Creek outcropping on nearby Black Creek (D. Prowell, personal communication to Albert E. Sanders, 2006), leaves little doubt that this is the unit exposed in the creek bank at this site (Figure 3.2). However, vertebrate material collected at the site evidently was reworked from the underlying Bladen Formation into a lag deposit at the base of the sampled outcrop (see discussion that follows).

Description and History

This site is located on a small, apparently unnamed creek about 0.1 mile (0.16 km) east of Sand Pit Road (County Road 1510; see Figure 1.1, p. 4) just east of the small town of Quinby, situated 1 mile (1.6 km) north of Florence. The creek empties into Black Creek approximately 0.45 mile (0.72 km) due west of the site. The streambed is no more than 5 feet (1.5 m) in width at its greatest breadth and traverses a wooded area along the foot of a high bluff. During the latter part of 1989, Derwin Hudson, Lee Hudson, Frank Morning, Jr., and Ray Ogilvie began to collect Cretaceous fossils from the bed of the creek and continued to do so until 1994. It was at this locality that the first dinosaur remains recorded from Florence County—ChM PV5400, PV6878 and SCSM 98.64.1—were discovered, the latter specimen found by Adam Perritt in 1993.

A dark, shaly, laminated clay unit, recognizably similar to portions of the Donoho Creek and Bladen Formations, is exposed on the sides of the creek near the water level. Frank Morning recalls that the specimens were collected from below that outcrop in a lag deposit that was exposed in the bed of the creek at that time but has subsequently been covered with sand by stream deposition. Considering the close proximity of this site to Black Creek and the pollen date from the outcrop at this site, this lag deposit could only have been the one that separates the Donoho Creek and Bladen Formations according to field observations by Prowell (D. Prowell, personal communication to Albert E. Sanders, 2006). Thus, the fossils from this lag deposit evidently were reworked from the underlying Bladen Formation.

Weishampel and Young (1996, 194) noted that "At least three sites near Quinby have yielded single theropod or hadrosaurid teeth," but those authors did not indicate the specific localities, the collectors, the number of specimens, or their whereabouts. Besides the site presently under discussion, the other two localities near Quinby are on TV Road (County Road 29) west of Quinby (F. Morning, personal communication to Albert E. Sanders, 2006), but those have not been included in the present review because the site under discussion has a better representation of material.

SITE NUMBER 3: TURBEVILLE

S.C., Clarendon County; spoil from pond on Richard Byrd property, south side of County Road 45 (Puddin Swamp Road), ca. 0.7 mile (1.13 km) south of U.S. Route 378 near Turbeville. USGS Turbeville 7.5′ quadrangle, 33° 52′ 49″N, 80° 03′ 08″W.

Formation and Age

Steel Creek Formation equivalent; late Maastrichtian, Late Cretaceous.

Determined from pollen extracted from augur-hole samples examined by R. Christopher and referred to SU (*Sparganeaepollinites uniformis*) Zone of pollen zonation scale. Although dinoflagellate cysts (marine phytoplankton) were present in the samples, the dominance of terrestrial organic materials (plant cuticle, wood fibers, pollen, spores) led Christopher to interpret the sampled sediments as having "been deposited in a proximal depositional setting (e.g., interdistributary bay)" (R. Christopher, personal communication to Albert E. Sanders, 2005).

Christopher also noted that the Peedee and the Steel Creek Formations "appear to be facies related, with the Peedee representing deposition in open marine environments, and the Steel Creek representing deposition in upper delta plain environments." Hence, the near-shore environment represented in the sediment samples from Site 3 is more nearly like that of the Steel Creek Formation than the open marine environment of the Peedee Formation.

Description and History

This site furnished the first known dinosaur specimens collected in South Carolina (ChM PV5400, PV6819), but they did not come to light until many years after they were found. About 1965, Mr. John C. Byrd excavated for a small pond at the rear of his property at 5324 Puddin Swamp Road, approximately 2.5 miles (4.0 km) southeast of the town of Turbeville (J.C. Byrd, personal communication to Albert E. Sanders, 1998). Sometime thereafter, he had the pond deepened and the spoil piles leveled out on the ground around it. Byrd soon began to find fossil bones in the spoil material. In 1995, his collection came to the attention of Frank Morning of Florence, and Morning expressed interest in the fossils in connection with the site investigations that he and his colleagues were conducting. Byrd generously gave several specimens to them, among which were two dinosaur specimens, one of which is reported here as a pedal phalanx (ChM PV6819; see Figure 4.27F, p. 99) similar in size to those of the tyrannosauroid *Appalachiosaurus montgomeriensis*, the second being a vertebral fragment (ChM PV 6777), possibly representing a hadrosaur.

With the permission of the present owner, Mr. Richard Byrd, an auger hole was put down near the pond at this site by R. Willoughby and others of the South Carolina Geological Survey, and Willoughby's drill log documented the following strata (all color references use the Munsell Color System, cited in Goddard et al. 1984):

0–5 feet (0–1.5 m): Moved earth from pond excavation

5–27 feet (1.5– 8.2 m): Duplin Formation (mid-Pliocene)

Upper Cretaceous sediments: 27–30 feet (8.2–9.1 m): Lightly cohesive brownish black (5YR 2/1) wet, moderately sorted, subangular to subrounded, very fine to medium quartz sand; with abundant colloidal(?) to very fine organic matter; locally to very coarse quartz sand; locally medium bluish red (5RP 5/1).

30–33 feet (9.1–10.1 m): Loose, wet, brownish gray (5YR 4/1), poorly to moderately sorted, subrounded, very fine to medium quartz sand, with minor clay; with black clay layer (2–4 cm thick) at 32 feet (9.75 m).

33–35 feet (10.1–10.7 m): Loose, light bluish gray (5B 4/1), calcareous, well sorted, subrounded, very fine to fine quartz sand, with scattered dark heavy minerals; slight fizz to dilute hydrochloric acid.

35 feet (10.7 m): Refusal. Blunted tip of drill bit.

SITE NUMBER 4: ASHLAND PAVING AND CONSTRUCTION COMPANY QUARRY NEAR LYNCHBURG

S.C., Lee County; Ashland Paving and Construction Company quarry, east side of Back Swamp Road (County Road 99), 1.45 miles (2.3 km) southeast of Durant Ferry Road (west fork of County Road 99) near Lynchburg.

USGS Lynchburg 7.5' quadrangle, 34° 03' 08"N, 80° 00' 57"W.

Formation and Age

Peedee Formation; early Maastrichtian, Late Cretaceous.

This locality is a now-defunct quarry formerly operated by the Ashland Paving and Construction Company. Allmon and Knight (1993, 356) reported ammonites and turritellid gastropods found in "hard, muddy to sandy limestone" blocks removed from the quarry and identified by them as the Peedee Formation based on the presence of the gastropod *Haustator bilira* and the ammonite *Sphenodiscus lobatus*, following Sohl and Owens (1991, figs. 11-6 and 11-15) who recorded *Haustator bilira* and *Sphenodiscus* species as marker species for the upper part of the calcareous portion of the Peedee Formation. Cretaceous vertebrate fossils were collected at this site in a stratum above the indurated bed that furnished the turritellids and ammonites. As at the Diamondhead Loop Road site (No. 7), the specimens clearly came from the noncalcareous sediments of the Peedee Formation overlying the calcareous portion at this locality.

Description and History

The vertebrate material was collected at this site by D. Hudson, F. Morning, and R. Ogilvie in 1988 and consisted mostly of dissociated crocodilian, turtle, and mosasaur fragments (F. Morning and R. Ogilvie, personal communication to Albert E. Sanders, 2006). The bulk of that material is in their respective private collections. Because of the large volume of specimens included in this volume, we have selected only one trionychid costal fragment (ChM PV6800) as a representative specimen from this locality. Weishampel and Young (1996, 194) stated that "a whole theropod tooth" had been found at "a locality" at Lynchburg in 1991 but gave no details as to the collector or the whereabouts of the specimen. It is not in the South Carolina State Museum or The Charleston Museum, and neither Morning nor Ogilvie are aware of any dinosaur material having been found at Lynchburg. Thus, it would appear that those authors were misinformed as to the correct locality at which the tooth was found.

SITE NUMBER 5: BLACK CREEK

S.C., Florence Co.; south bank of Black Creek, c. 0.215 mile (0.345 km) southwest of County Road 343 (TV Road) near Quinby.

USGS Florence East 7.5' quadrangle, 34° 14' 29" N, 79° 44' 35" W.

Formation and Age

Donoho Creek Formation; early late Campanian, Late Cretaceous.

Description and History

As noted previously, Black Creek in Darlington and Florence counties, South Carolina, is the type of locality described in the "Black Creek phase" of Sloan (1908), later termed the Black Creek Formation by Stephenson (1912). More recent interpretations (Owens 1989; Sohl and Owens 1991) elevated the Black Creek Formation to Group status. Two formations in the Black Creek Group, the Bladen and the overlying Donoho Creek, occur along Black Creek in Darlington and Florence counties and are separated by a lag deposit.

In 2003 Ray Ogilvie collected a parietal bone of an indeterminate eusuchian crocodilian (ChM PV8848) from the Donoho Creek Formation at an outcrop on the south bank of Black Creek approximately 0.215 miles (0.345 km) southwest of TV Road (County Road 343) near Quinby in Florence County. That specimen is the only one from the Black Creek-type locality that has come to our attention.

SITE NUMBER 6: MULDROW'S MILL

S.C., Florence Co.; drainage ditches ca. 0.10 mile (1.6 km) west of U.S. Routes 52 and 301 near Muldrow's Mill, ca. 0.2 mile (0.32 km) north-northeast of County Road 107 in Florence.

USGS Florence West 7.5' quadrangle, 34° 08' 13" N, 79° 46' 21" W.

Formation and Age

Peedee Formation, early Maastrichtian, Late Cretaceous.

Assigning the Cretaceous fossil-bearing beds at this site to the Peedee Formation is based on Frank Morning's observation of the stratigraphy in a sand pit on nearby Bannockburn Road (County Road 722) approximately 0.7 mile (1.12 km) southeast of Muldrow's Mill. Quarrying operations in the pit exposed the indurated calcareous lower portion of the Peedee Formation underlying unconsolidated detrital sediments that furnished Cretaceous vertebrate fossils similar to those collected from the Muldrow's Mill site. The vertebrate fossil-bearing sediments at both sites are lithologically similar and both contain specimens of *Belemnitella*. Thus, there is little doubt that the vertebrate material from both the Muldrow's Mill and Bannockburn Road sites was collected from the detrital upper portion of the Peedee as observed at Burches Ferry by Self-Trail et al. (2002) and that is apparently present also at Site Number 7 as discussed in the text that follows.

Description and History

This relatively small site consisted of an L-shaped ditch approximately one-tenth of a mile (200 m) west of U.S. Routes 52 and 301 near Muldrow's Mill, an historic site in the southern portion of the city of Florence (F. Morning, personal communication to Albert E. Sanders, 2006). Cretaceous vertebrate remains, primarily isolated, fragmentary bones or teeth of crocodilians, turtles, mosasaurs, and fishes, were collected from the ditches by D. Hudson, F. Morning, and R. Ogilvie in 1989 and are in their private collections. We have chosen to include only three fragments of crocodilian osteoderms (ChM PV8849–PV8851) as representative material from this locality.

SITE NUMBER 7: DIAMONDHEAD LOOP ROAD

S.C., Florence County; pond on south side of Diamondhead Loop Road; ca. 1.3 miles (2.1 km) south of Claussen via South Carolina Route 327, thence 1.4 miles (2.3 km) left (south) on Diamondhead Loop Road, thence 0.2 mile(0.32 km) right (south) on dirt road.

USGS Pamplico North 7.5′ quadrangle, 34° 05′ 52″ N, 79 ° 37′ 16″ W.

Formation and Age

Peedee Formation, early Maastrichtian, Late Cretaceous.

Description and History

This site was discovered by F. Morning in 2000, the result of an excavation for a small pond approximately 0.2 mile (0.32 km) south of Diamondhead Loop Road and ca. 0.3 mile (0.48 km) NE of Willow Creek in eastern Florence County. Cretaceous fossils were found in the spoil material from the excavation, but since that time the margins of the pond have been covered with grass and other vegetation, thus precluding the collection of matrix samples for analysis. However, Morning collected the specimens from this site from unconsolidated sediments just above the indurated Peedee, which places them in the detrital upper portion of that unit as it occurs at Burches Ferry, described by Self-Trail et al. (2002, 147) as "noncalcareous with mottled coloring, varying between light olive gray and moderate yellowish brown." That is in keeping with the yellowish-brown coloration of some of the fossils found at this site. Many of the specimens from this locality were badly worn and largely indeterminate, but we have included two specimens as documentation of the site, one, an indeterminate mosasaur tooth (ChM PV7312), and the other a coprolite, probably crocodilian (PV8997).

SITE NUMBER 8: BURCHES FERRY

S.C., Florence County; Burches Ferry, west bank of Pee Dee River, ca. 0.75 mile (1.21 km) east of County Road 52, ca. 0.08 mile (0.18 km) northwest of junction of Mill Branch and Pee Dee River.

USGS Pamplico North 7.5' quadrangle, 34° 03' 50" N, 79° 31' 43" W.

Formation and Age

Donoho Creek Formation, early late Campanian; Peedee Formation, early Maastrichtian, Late Cretaceous.

Description and History

As discussed previously, the Cretaceous beds at Burches Ferry and other sites on the Pee Dee River are among the earliest studied in the eastern United States, having been discussed extensively by Ruffin (1843) and Tuomey (1848). This classic locality affords excellent exposures of the Donoho Creek and Peedee Formations (Figure 3.3) and was designated as the type locality for the Peedee Formation by Stephenson (1923). The Peedee is also the source of the Peedee belemnite (PDB), which is used as an international standard for paleotemperature analyses, based on the oxygen and carbon isotopic ratios in the calcite of the phragmocone.

A corehole (FLO-311) put down by the U.S. Geological Survey produced important information about the stratigraphy at this locality, clarifying previous debate about the contact between the Peedee and the underlying Donoho Creek Formation at Burches Ferry (Benson 1969; Van Nieuwenhuise and Kanes 1976; Lawrence and Hall 1987; Sohl and Owens 1991). Drilled to a depth of 44 feet, the core contained 23.7 feet of Donoho Creek Formation overlain by 15 feet of Peedee Formation and 5.3 feet of Quaternary sand (Self-Trail et al., 2002). "The basal 2.5 feet of the Peedee Formation is clayey sand containing rounded phosphate pebbles and clasts. . . . Vertebrate remains are concentrated in this lag deposit, as are pieces of permineralized wood up to 0.5 inches in length" (Self-Trail et al. 2002, 147).

Figure 3.3 Section at Burches Ferry on the Pee Dee River in Florence County, South Carolina, showing outcrops of Campanian Donoho Creek Formation and Maastrichtian Peedee Formation (Site 8, Figure 1.1).

Examination of both the outcrop and the core-hole material led Self-Trail et al. (2002, 154, 157) to conclude that "the Peedee Formation exposed at its type locality, Burches Ferry, consists of massively reworked upper Campanian sediments" of the Donoho Creek Formation. Analysis of the calcareous nannoplankton showed that the Donoho Creek strata at this locality are referable to nannoplankton Zone CC22c, and also revealed that the nannofossils in the Peedee Formation at Burches Ferry are massively reworked (that is, eroded and redeposited, possibly more than once). Nevertheless, calcareous nannoplankton from Peedee sediments in the core are indicative of Zone CC25a (Self-Trail et al. 2002). The complete absence of the Zone CC23 and Zone CC24 beds at Burches Ferry suggests that they were eroded by the first marine transgression forming the Peedee Formation (Self-Trail et al. 2002), and their absence has disclosed a gap of about five million years in the chronostratigraphic record of late Campanian and early Maastrichtian deposition in South

Carolina (J. Self-Trail, personal communication to Albert E. Sanders, 2006; see Table 1.1, p. 6).

Vertebrate fossils from Burches Ferry have come predominately from the Donoho Creek Formation, perhaps because of the greater exposure of that unit at this locality, especially at times when the water levels of the river are low. Weishampel and Young (1996, 194) stated that "This site . . . is seldom exposed because of a strictly controlled flow schedule from dams along the Pee Dee River," but those changes in the level of the river mostly affect the exposure of the lower portion of the Donoho Creek Formation. The upper portion of that unit and the overlying Peedee Formation are usually accessible. Cicimurri (2007) has reported 22 species of sharks and rays from the Donoho Creek Formation at Burches Ferry.

Specimens from Burches Ferry used in the present study were collected by Curtis Bentley, Mike Bruggeman, Bryan England, and Lee Hudson.

SITE NUMBER 9: PEE DEE RIVER NEAR ALLISON FERRY LANDING

S.C., Florence Co.; west bank of Pee Dee River ca. 0.25 mile (0.40 km) south of Allison Ferry Landing, ca. 1.25 miles (2.01 km) northeast of Poston.

USGS Gresham 7.5′ quadrangle, 33° 52′ 53″N, 79° 24′ 24″W.

Formation and Age

Donoho Creek Formation; latest middle–earliest late Campanian, Late Cretaceous.

Description and History

A short distance downstream from the site of the historic Allison Ferry, this locality has furnished three well-preserved associated vertebrae of a plesiosaur (ChM PV7181; see Figure 4.12A–C, p. 70). Collected by Bryan England and Abigail Pfaff, the specimens were found in the Donoho

Creek Formation close to each other and evidently belong to a single individual. They are discussed in detail in Chapter 4

SITE NUMBER 10: PEE DEE RIVER, TWO MILES SOUTH OF ALLISON FERRY LANDING

S.C., Florence Co.; west bank of Pee Dee River, two miles (3.2 km) south of Allison Ferry Landing.

USGS Johnsonville 7.5′ quadrangle, 33°51′ 24″ N, 79° 22′ 51″ W.

Formation and Age

Peedee Formation; early Maastrichtian, Late Cretaceous.

Description and History

This site furnished one specimen, a posterior peripheral (ChM PV8596) of the sea turtle *Peritresius ornatus* (see Figure 4.9 A–C, p. 64), collected from the Peedee Formation by Bryan England in the fall of 2001.

SITE NUMBER 11: CLAPP CREEK, KINGSTREE

S.C. Williamsburg County: Kingstree, along Clapp Creek, east side of the town of Kingstree.

USGS Kingstree 7.5′ quadrangle. Coordinates not determined.

Formation and Age

Donoho Creek Formation, early late Campanian, Late Cretaceous.

Within and near the town of Kingstree, sediments of the Williamsburg Formation (Black Mingo Group, late Paleocene) occur just beneath the surface along the banks of Clapp Creek. Less than one meter thick, this

Paleocene deposit is underlain by dark-gray, laminated, silty clay of the Late Cretaceous Donoho Creek Formation. Stratigraphic definition of these beds is poorly understood, however, as taxa of mixed ages ranging from Late Cretaceous to Pliocene have been reported from these deposits (Erickson 1998; Weems and Bybell 1998). Some of this material appears to have been transported from some distance; but the greatest concentration of bone fragments comes from the uppermost sediments of the late Paleocene Williamsburg Formation. Specimens from the underlying Cretaceous sediments are less numerous but better preserved, suggesting reworking of those specimens.

Description and History

Dinosaur remains consisting of two hadrosaurid teeth were collected here by Aura Baker in 1986 (Weishampel and Young 1996). A third hadrosaurid tooth fragment was recovered in 1987, and a fourth in 2006. During 1984 and 1985, Bruce Erickson and Bruce Lampright collected numerous specimens at this location from the sides and bottom of Clapp Creek by hand, shovel, and sieve. Among them is an incomplete left exoccipital (SMM P84.12.27) of a hadrosaur that is the earliest collected dinosaur specimen from Kingstree in the assemblage. By excavating pits at various intervals for a distance of about 20 meters on either side of the creek both Paleocene and Cretaceous levels were sampled. Individual pits were dug through Paleocene sediments below the level of the bottom of the creek bed to reach the Cretaceous deposits. Materials from both levels were separately concentrated by wet-sieving the matrix through graduated mesh sizes. During this process stream currents were barely sufficient to remove the sediments from the sieve boxes, therefore, hand selection of specimens was necessary (Figure 3.4). Many specimens from both levels were removed by this method.

Subsequently, a total of nine additional fragments of dinosaur bones and teeth and a tooth of the giant crocodilian *Deinosuchus* have been found at this locality and placed in the collections of The Charleston Museum and the South Carolina State Museum.

Figure 3.4 The Campanian Donoho Creek Formation occurs as a silty clay beneath late Paleocene deposits along Clapp Creek at Kingstree, Williamsburg County, South Carolina (Site 11, Figure 1.1). Specimens collected involved wet-sieving matrix.

SITE N0. 12: WACCAMAW RIVER

S.C., Horry County; bottom of Waccamaw River at Wild Horse subdivision, ca. 6 miles (9.6 km) east of Conway.

Coordinates not determined.

Formation and Age

Formation undetermined; Late Cretaceous.

Description and History

A large mosasaur tooth (SCSM 94.107.1) collected at the bottom of the Waccamaw River six miles (9.6 km) east of Conway by Ron Willman of

Conway in 1994 documents the occurrence of Late Cretaceous vertebrates at this locality. However, the specimen was not accompanied by sediment samples that would indicate the horizon from which this specimen originated.

SITE NO. 13: MYRTLE BEACH

S.C., Horry County; Myrtle Beach (no further data).

Coordinates could not be determined.

Formation and Age

?Peedee Formation; ?early Maastrichtian, Late Cretaceous.

The presence of outcrops of the Peedee Formation in the banks of the Intracoastal Waterway near Myrtle Beach suggests that the specimen recorded may have come came from that unit but not necessarily from that locality.

Description and History

The South Carolina State Museum contains a mosasaur tooth (SCSM 84.176.1) collected at Myrtle Beach by Mrs. E. A. Affinito and donated to the State Museum in 1984. There is no information regarding the date of collection or the specific location at which the specimen was found.

SITE NO. 14: NEAR LITTLE RIVER

S.C., Horry County; borrow pit beside U.S. Route 17 near town of Little River USGS Calabash (NC, SC) 7.5′ quadrangle.

Coordinates could not be determined.

Formation and Age

Donoho Creek Formation; ?early late Campanian, Late Cretaceous.

Description and History

The site is documented by one specimen, a vertebra (ChM 4818) of an indeterminate dinosaur found by Buster Burke in 1991. The specimen was donated to the ChM by the youngster, who said that he had found it in a pit beside U.S. Route 17 near Little River, a small town in Horry County located 1.9 miles (3.06 km) southeast of the North Carolina state line. Recognizing the specimen as a probable dinosaur vertebra, Erickson and Sanders visited the Little River area in search of the roadside excavation but were unable to determine its whereabouts with certainty. Hence, the stratigraphic origin of the specimen is uncertain, but it is probably of Campanian age and most likely came from a lag deposit at the base of a Plio-Pleistocene unit.

4

SYSTEMATIC PALEONTOLOGY

INSTITUTIONAL ABBREVIATIONS

ChM, PV, Charleston Museum, Charleston, South Carolina
NCSM, North Carolina Museum of Natural Sciences
NJSM, New Jersey State Museum
SCSM, South Carolina State Museum, Columbia, South Carolina
SMM, The Science Museum of Minnesota, St. Paul, Minnesota.

Class REPTILIA Linneaus, 1758
Order TESTUDINES Linneaus, 1758; Batsch, 1788
Suborder PLEURODIRA Cope, 1864
Family BOTHREMYDIDAE Baur, 1891
Tribe BOTHREMYDINAE Gaffney, Tong and Meylan, 2006
Bothremys Leidy, 1865
Bothremys sp.
Figure 4.1

MATERIAL

Site 1, Stokes Quarry–ChM PV8702, right mandible.

DISCUSSION

ChM PV8702 is a complete right mandible of an undetermined species of *Bothremys*. It is an elongate jaw (100.4 mm) with an elongate symphysis (30.2 mm). The dentary is dorsoventrally thick with a broad, concave, triturating surface. A ventral tongue of the dentary reaches posteriorly to the level of the articular, where it is overlain by the surangular labially and by the angular lingually. The lingual margin of the dentary is upturned forming a high ridge between the top of the coronoid process and descending to the anterior part of the symphysis. The cone-shaped pit characteristic of *Bothremys* is well defined (Gaffney, Tong, and Meylan 2006). The coronoid process is situated at mid-length of the jaw. It is tall with a

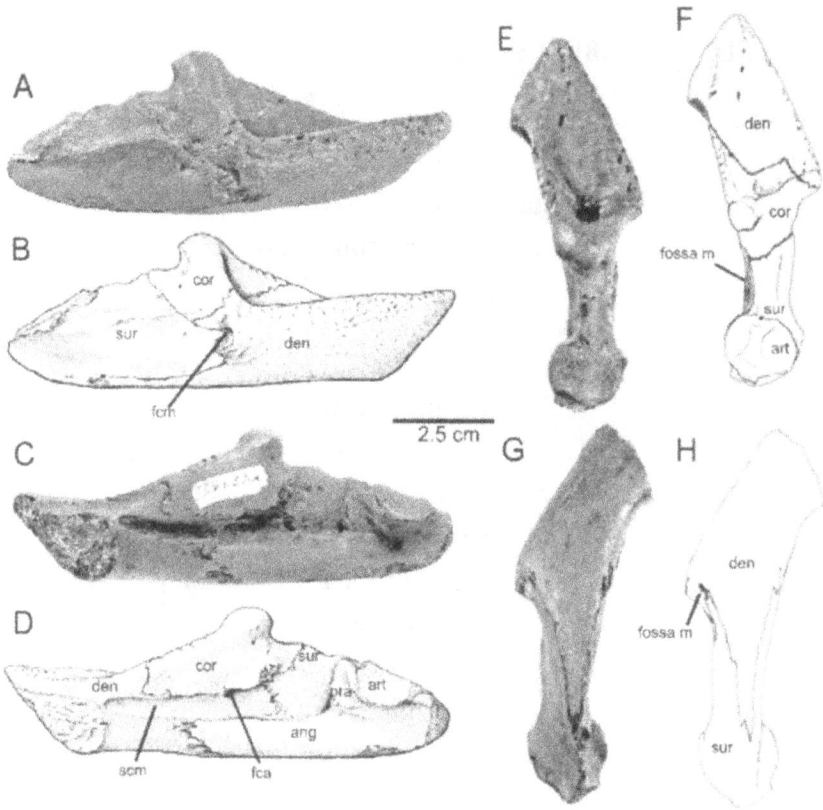

Figure 4.1 Right mandible, *Bothremys* sp. indet. (ChM PV 8702). (A, B): Right lateral external view. (C, D): Right lateral internal view. (E, F): Dorsal view. (G, H): Ventral view. Abbreviations: **ang** = angular, **art** = articular, **cor** = coronoid, **den** = dentary, **fai** = foramen alveolare inferius, **fca** = foramen cavalis aloedaris, **fcm** = foramen canalis mentale, **fossa m** = foramen canalis mentale, **fossa scm** = sulcus cartilaginis meckelii, **pra** = prearticular; **sur** = surangular. Scale bar = 2.5 cm.

domed top having a slight posterior overhang (Figure 4.1, especially A–D). The dorsal half of the lingual ridge is provided by the coronoid. Functionally, the coronoid and surangular provide surface areas labially for insertion of the mandibular adductor (par media), which originates from the anterior surface of the quadrate and runs to the posterior edge of the coronoid process (Schumacher 1973). As seen in the measurements that follow, the surangular makes up over half the mandibular length. The anterior portion of the prearticular above the Meckelian sulcus is missing and there is no indication of a splenial. The Meckelian sulcus ends short of the mandibular symphysis at the symphysial concavity.

ChM PV8702 is identified as *Bothremys* because the dorsal pit of the triturating surface is present, though it is less prominent than in NCSM 14499, a partial left ramus of *Bothremys* sp. from the upper Campanian of North Carolina reviewed by Gaffney, Hooks, and Schneider (2009, fig. 9). The new mandible also has a symphysial concavity that is shallower than that of the *Bothremys* presented by Gaffney, Hooks, and Schneider (2009, fig. 9). Whether these differences are taxonomically significant or whether they simply reflect individual variation is presently indeterminate. Measurements of the specimen in millimeters are as follows:

Anteroposterior length of mandible.. 100.4
Height of mandible at coronoid process... 35.0
Length of dentary 80.6
Breadth of dentary (including dentofacial process) 28.0
Breadth of coronoid.. 25.0
Thickness of dentary at midlength 15.0
Ventral overlap of dentary/angular... 46.0
Length of triturating surface 40.2
Length of symphysis.. 30.2
Height of symphysis (posterior) 10.3
Height of symphysis (anterior) 30.2
Length of angular 64.3
Length of surangular 64.2
Breadth across articular mandibularis... 23.0

Bothremys Leidy, 1865 or *Chedighaii* Gaffney, Tong and Meylan, 2006
Bothremys or Chedighaii sp
Figure 4.2

MATERIAL

Site 1, Stokes Quarry—ChM PV8693 and PV8694, xiphiplastron fragments; PV8902, fragment of hypoplastron; PV8905, plastral fragment;

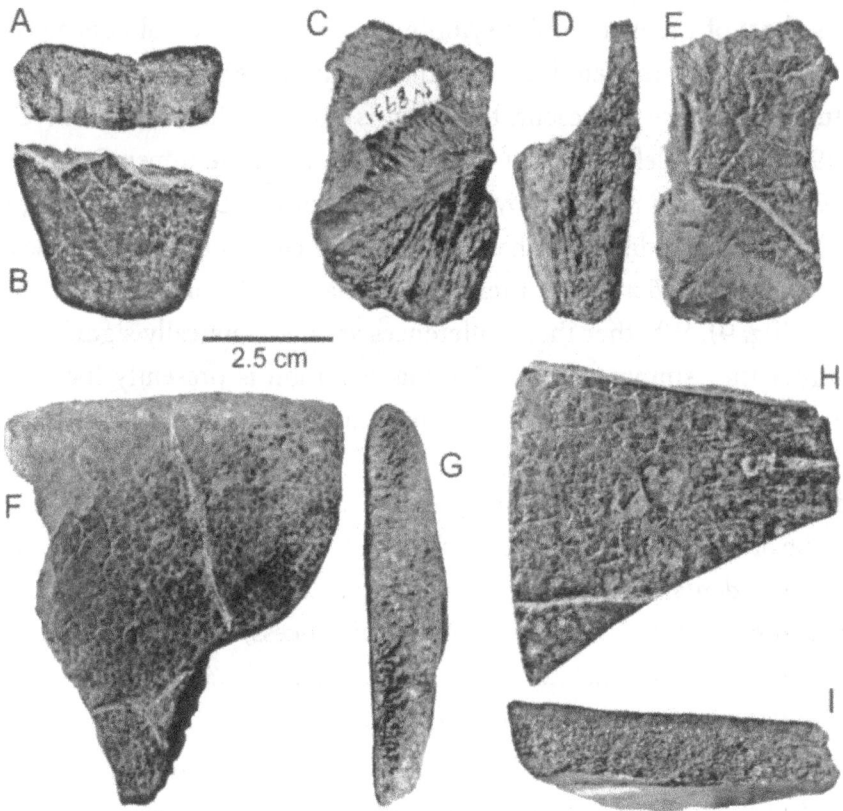

Figure 4.2 Carapace and plastron fragments of *Bothremys*, *Chedighaii*, or both. (A, B): ChM Pv8932, posterior portion of neural in cross-sectional and dorsal views. (C, D, E): PV8931, anterior portion of right xiphiplastron in internal view showing ilial scar, cross-sectional view, and dorsal view with sulcus. (F, G): PV8926, first left peripheral in dorsal and anterior views. (H, I): PV8928, fragment of plastron with midline suture (right) and sulcal groove in external (ventral) view and posterior cross-sectional view. Scale bar = 2.5 cm.

PV8909, peripheral fragment; PV8913, PV8916, costal fragments; PV8924, right xiphiplastron fragment; PV8926, first left peripheral; PV8927, costal 7 fragment; PV8928, plastral fragment; PV8929, proximal end of left costal; PV8931, left xiphiplastron fragment; PV8932, posterior part of neural 3 or 4; SMM P2006.1.8 and P2006.1.9, costal fragments.

DISCUSSION

The carapace and plastron fragments illustrated here (Figure 4.2) have an external sculpture that is characteristic of bothremydine turtles, but it is impossible presently to determine whether they belong to *Bothremys*, *Chedighaii*, or represent a mixture of both. Both genera were defined on skull and jaw characteristics in the absence of any knowledge of their correlative postcranial remains (Gaffney et al. 2006), so it is not at all clear whether these shell fragments pertain to one or both of these genera. Both genera have been recognized from upper Campanian strata in southern North Carolina based on skull and jaw fragments (Gaffney et al. 2009), so both genera are expected to be from the South Carolina upper Campanian beds that yielded the bothremydine postcranial remains described here.

Carapace fragments include ChM PV8932 (see Figure 4.2A–B), which is the posterior half of neural element 3 or 4, and ChM PV8926 (Figure 4.2F–G), which is a second left peripheral. This second left peripheral is very similar to the same element in *Bothremys* (or *Chedighaii*) *barberi* (Gaffney et al. 2006, 554). Plastral elements include ChM PV8928 (see Figure 4.2H–I), which is a fragment of plastron with a midline suture (on the right) and a sulcal groove in external (ventral) view, and PV8931 (see Figure 4.2C–E), which is a xiphiplastron fragment that possesses a prominent pubic scar on its internal surface and preserves a section of the femoral/anal sulcus on its external surface.

Suborder CRYPTODIRA Cope, 1868
Infraorder PARACRYPTODIRA Gaffney, 1975a
Family MACROBAENIDAE Sukhanov, 1964
Osteopygis Cope, 1869a
Osteopygis emarginatus Cope, 1869a
Figure 4.3G–I

MATERIAL

Site 1, Stokes Quarry—ChM PV8912, tenth left peripheral.

DISCUSSION

A posterior peripheral ChM PV8912 (Figure 4.3G–I) is the only specimen that can be confidently assigned to this taxon. The internal margin of

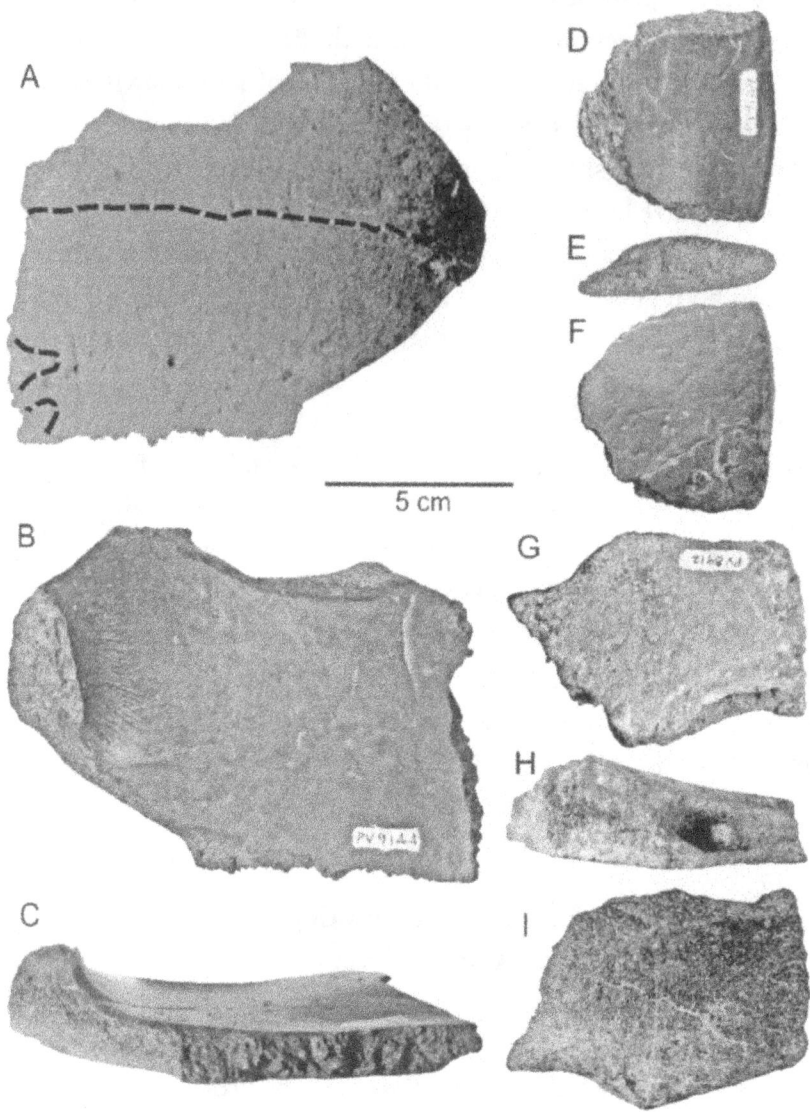

Figure 4.3. Carapace fragments of *Adocus* cf. *A. punctatus* and *Osteopygis emarginatus*. (**A, B, C**): ChM PV9144, left hyoplastral fragment of *Adocus* in external (*ventral*), internal (*dorsal*), and anterior views. (**D, E, F**): PV9129, first right peripheral of *Adocus* in ventral, anterior, and dorsal views. (**G, H, I**): PV8912, large fragment of tenth left peripheral of *Osteopygis emarginatus* in ventral, internal, and dorsal views. Scale bar = 5 cm.

this element is somewhat worn, so it is not clear whether a costoperipheral fontanelle was present or not. However, the rather flat dorsal surface of this element, the presence of a very deep socket for the rib of the adjacent costal element, and the presence of an anteroposterior sulcal groove near the internal margin of this peripheral (which once separated a pleural scute from the adjacent marginal scute) together are characteristics typical of this taxon. In most chelonioid turtles, and in *Adocus*, the anteroposterior sulcus lies closer to the midline than the peripheral border in posterior peripherals. In *Agomphus*, the posterior peripherals are strongly upturned concavely, so this element does not appear to belong to any of these taxa. The lack of any obvious pattern on the external surface precludes assignment of this specimen to any of the other taxa discussed here. See also the discussion to follow for *Euclastes*, concerning assignment of specimens here to *Osteopygis*.

Infraorder EUCRYPTODIRA Gaffney, 1975a
Superfamily TRIONYCHIA Hummel, 1929
Family ADOCIDAE Cope, 1870
Adocus Cope, 1868
Adocus cf. *A. punctatus*
Figures 4.3A–F, 4.4

MATERIAL

Site 1, Stokes Quarry—ChM PV8903, hypoplastron fragment, PV8925, plastron fragment, PV8930, costal fragment, PV9129, first right peripheral; PV9135, bridge fragment; PV9144, left hypoplastron fragment; SMM P2006.1.1, fifth neural.
Site 8, Burches Ferry—SCSM 2005.25.1, section of center carapace including all or parts of neurals and proximal costals three, four, and five; SCSM 2005.25.12, costal fragment; SCSM 2005.25.21, neural.

DISCUSSION

When unworn, the carapace and plastron of smaller specimens of *Adocus* are ornamented by fine shallow pits. In larger and more worn specimens, this pitting may be faint or even absent (Hay 1908; Hutchison and Weems

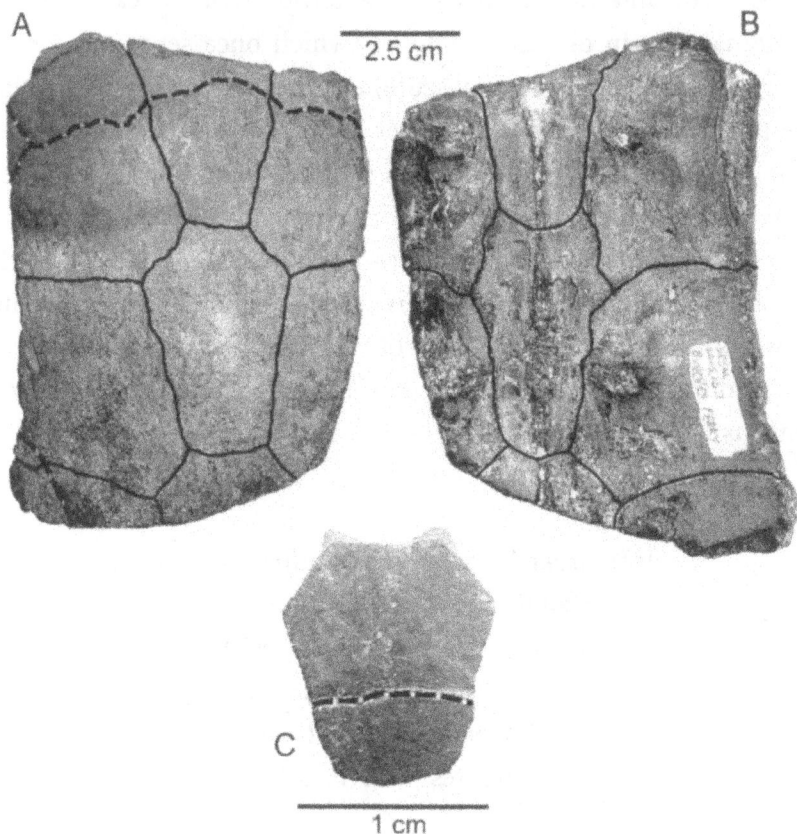

Figure 4.4 Carapace of *Adocus* cf. *A. punctatus*. (A, B): Section of carapace (SCSM 2005.25.1) in external and internal views, including all or parts of neurals and proximal costals three, four, and five. (C): SMM P2006.1.1, neural 5, external view. Scale bar = 5 cm.

1998). One of the few specimens in the assemblage that has several bones in articulation (SCSM 2005.25.1, Figure 4.4A–B) preserves joined neurals three, four, and five and the proximal parts of costals three, four, and five. The sulci indicating the anterior and posterolateral margins of the second neural scute are also preserved in this specimen. Surface pitting is present but mostly obscured perhaps due to the advanced age of this individual. On the internal surface all sutures are sharply defined and the rib heads are vestigial as is typical of *Adocus* (Hay 1908). SMM P2006.1.1 (Figure 4.4C) is a complete fifth neural with a sulcus indicating the contact between vertebral scutes two and three, as in *Adocus*. Its dorsal surface lacks pitting, but as mentioned previously the diagnostic shell texturing is not always present. A large fragment of the left hypoplastron (ChM PV9144, see Figure 4.3A–C) has features of *Adocus punctatus*, including a straight, narrow, shallow sulcus marking the transverse contact of the abdominal and femoral scutes; irregular median sulci that wander across the midline suture; and fine almost indistinguishable pitting on the ventral surface. These are all characteristic of *Adocus* and not *Agomphus*. Specimens referable to *Adocus* are fairly abundant, but most are not complete enough for species identification. As the hypoplastron fragment is similar to *A. punctatus*, this material is tentatively assigned to that taxon. It is likely but not certain that only one species is represented.

Family Trionychidae Gray, 1825
Genus indeterminate
"Trionyx" halophilus Cope, 1869b
Figure 4.5A–R

MATERIAL

Site 1, Stokes Quarry—ChM PV8935, hyoplastral fragments; PV8936, PV8938, PV8941, PV8942, PV8943, shell fragments; PV8946, PV8947, shell fragments; PV8949, PV8950, PV8951, PV8952, shell fragments; PV8953, distal costal fragment; PV8954, costal fragment; PV8955, neural; PV9128, costal fragment; SMM P2600.1.2, hypoplastral fragment.
Site 2, Quinby—ChM PV6782, PV6783, PV6784, PV6785, PV6786, PV6787, PV6788, PV6789, PV6790, PV6791, PV6793, PV6795, PV6797, PV6798, costal fragments; PV6796, distal neural fragment.
Site 4 Lynchburg—PV6800, PV6801, PV6802, PV6803, PV6804, PV6805, PV6806, PV6807, PV6808, PV6809, PV6810, PV6811, PV6812, PV6813, PV6816, costal fragments.

DISCUSSION

These carapace and plastron fragments pertain to the Trionychidae because of their characteristic sculpturing, lack of sulci, and absence of

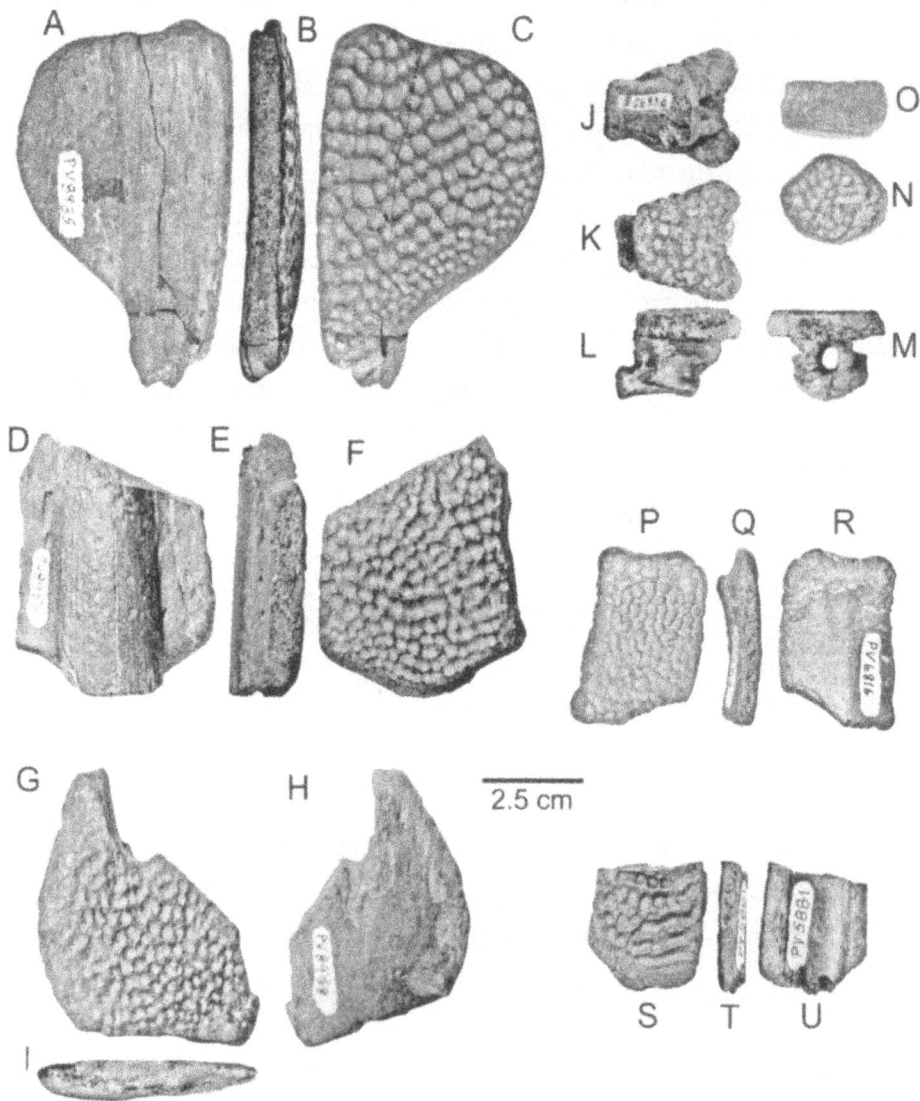

Figure 4.5. Carapace and plastron of "*Trionyx.*" (A, B, C): "*T.*" *halophilus*, ChM PV8935, distal portion of eighth left costal in ventral, anterior, and dorsal views. (D, E, F): "*T.*" *halophilus*, PV8953, distal portion of costal in ventral, lateral, and dorsal views. (G, H, I): "*T.*" *priscus*, PV8948, proximal end of right xiphiplastron in dorsal, ventral, and cross-sectional views. (J, K, L, M): "*T.*" *halophilus*, PV8955, neural and underlying vertebra in ventral, dorsal, right lateral, and anterior views. (N, O): PV6796, "*T.*" *halophilus*, posterior neural in dorsal and anterior views. (P, Q, R): "*T.*" *priscus*, PV6816, proximal costal in dorsal, lateral, and ventral views. (S, T, U): "*T.*" *priscus*, PV5881, distal costal in dorsal, lateral, and ventral views.

any evidence of peripheral elements (Hay 1908; Meylan 1987). Baird (1986) and Hutchison and Weems (1998) have noted the ambiguity of identifications of ontogenetic and individual variation based on fragmentary material, but at least two quite different patterns are present. As is true with most Paleogene turtle faunas of the eastern coastal United States, Upper Cretaceous turtles are found in near-shore marine sediments as part of mixed freshwater and marine taxa assemblages. It is possible that the trionychid material was washed into a shallow marine setting, but the abundance of trionychid material suggests that these species probably were salt-water tolerant and habitually moved into and out of the shallow marine environment.

The material described in this section can be assigned to the Upper Cretaceous taxon "*Trionyx*" *halophilus* (Cope 1869a). The type material of "*T.*" *halophilus* was found in Maastrichtian sediments in Delaware (Baird and Galton 1981), so the South Carolina material is in part slightly older but otherwise indistinguishable. Although *Trionyx* was once widely used as a generic designation for American Cenozoic soft-shell turtle specimens, the name does not properly apply to any American material (Meylan 1987). There are no characteristics preserved in the material here ascribed to "*Trionyx*" *halophilus* that could clearly establish what the correct generic assignment for this material should be, though the trionychid genus *Aspideretoides* (Gardner, Russell, and Brinkman 1995) of about the same age in western North America is a plausible possibility. For now, until better material is recovered, the established name "*Trionyx*" *halophilus* is retained with *Trionyx* in quotation marks for nomenclatural stability. Most of the material comes from Campanian strata, but two specimens from Muldrow's Mill are from the Maastrichtian Peedee Formation.

The external surfaces of these fragments are covered by an irregular pattern of rounded, more or less circular pits. One proximal costal fragment from a small individual (Figure 4.5P–R) is slightly different in that the pits are not so uniformly rounded as in larger specimens; it could represent a different taxon but more probably represents a juvenile pattern of "*T.*" *halophilus*. A variety of shell elements are represented, including distal portions of costals (ChM PV8935, PV8955, Figure 4.5J–M), the proximal portion of a costal (PV6816; Figure 4.5P–R), neurals (PV6796,

PV8955; Figure 4.5J–M), and the proximal end of a right xiphiplastron (PV8948; Figure 4.5 G–I). The elements are of average trionychid thickness.

"*Trionyx*" *priscus* Leidy, 1851
Figures 4.5 G–I, S–U, 4.6

MATERIAL

Site 1, Stokes Quarry—ChM P5881, costal fragment; PV8899, carapace fragment; PV8900, right hypoplastron; PV8937, PV8939, PV8940, shell fragments; PV8944, right hyoplastal fragment; PV8945, costal fragment; PV8948, xiphiplastral fragment; PV8957, costal fragment; PV9145, costal fragment; SMM P2600.1.2, hypoplastral fragment.
Site 2, Quinby—ChM PV6792, PV6794, shell fragments; PV6814, PV6815, PV6817, PV6818, costal fragments.

DISCUSSION

The fragments referable to this species differ from "*T.*" *halophilus* in that they have a more irregular pitting pattern and less continuous rims around the pits. Also, the distal ends of the costals have a strongly linear, closely spaced ridge-and-trough pattern (Figure 4.6C) that is not present on the distal costals of "*T.*" *halophilus* (see Figure 4.5C, F). A number of elements are represented, including a proximal costal fragment (ChM PV8957; Figure 4.6A–C), a medial costal fragment (PV8939; Figure 4.6D–F), two distal costal fragments (PV5881; see Figure 4.5S–U and PV9145; Figure 4.6J–L), a right hyoplastron fragment (PV8944, Figure 12G-I), and a right hypoplastron fragment (PV8900; Figure 4.6M–O). The costals of "*T.*" *priscus* are somewhat thicker distally than costals of similar size of "*T.*" *halophilus*. The type of "*T.*" *priscus* came from the Upper Cretaceous greensands of New Jersey. It is the only other species of trionychid besides "*T.*" *halophilus* that has been described from the Upper Cretaceous of the southeastern United States. Like "*T.*" *halophilus*, "*T.*" *priscus* cannot be

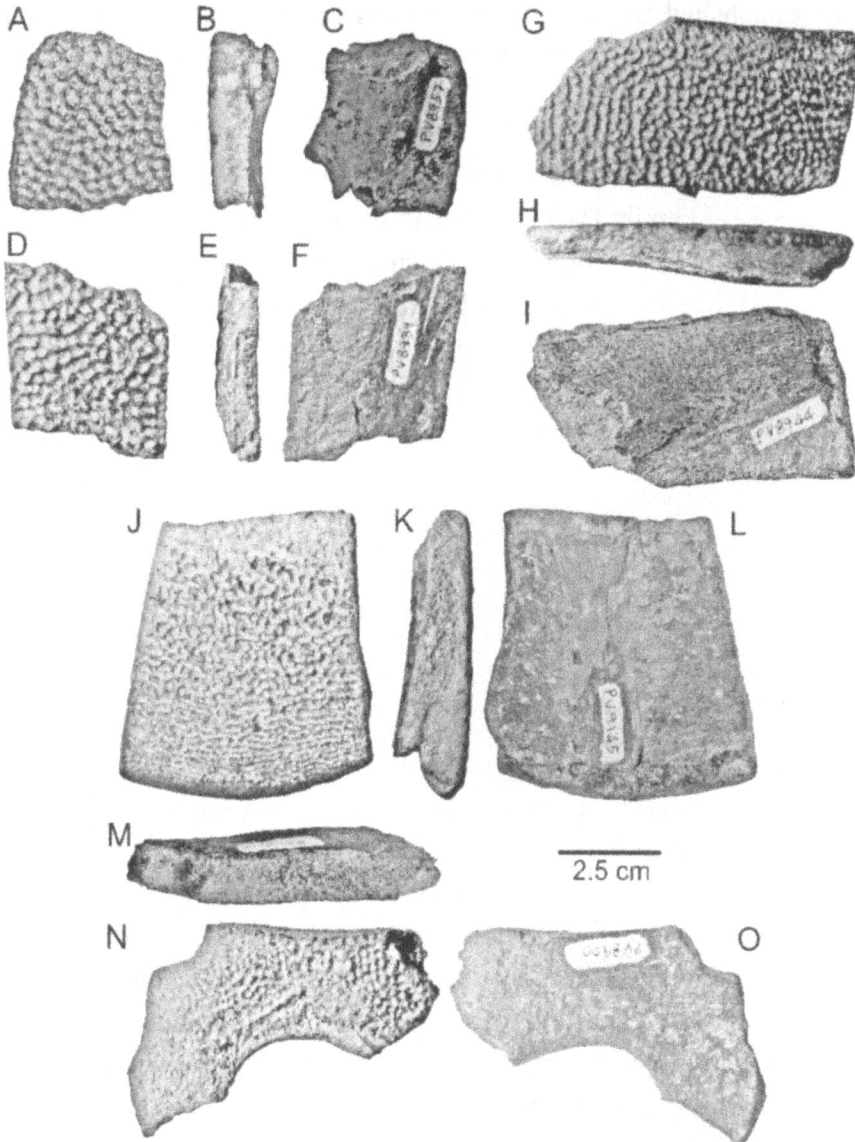

Figure 4.6 Carapace and plastron of *"Trionyx" priscus*. (A, B, C): ChM PV8957, proximal costal in dorsal, lateral, and ventral views. (D, E, F): PV8939, central region of costal in dorsal, lateral, and ventral views. (G, H, I): PV8944, anterior distal portion of right hyoplastron in dorsal, lateral, and ventral views. (J, K, L): PV9145, distal end of costal in dorsal, lateral, and ventral views. (M, N, O): PV8900, right hypoplastron in anterior, external (ventral), and internal (dorsal) views.

assigned with any certainty to any described genus of trionychid turtle, so the established name "*Trionyx*" *priscus* is retained with *Trionyx* in quotes for nomenclatural stability.

Superfamily CHELONIOIDEA
Family DERMOCHELYIDAE Fitzinger, 1843
Corsochelys Zangerl, 1960
Corsochelys bentleyi sp. nov.
Figures 4.7, 4.8

HOLOTYPE

SCSM 2005.25.3, articulated frontals and parietals forming a single skull roof.

DIAGNOSIS

This species belongs in the genus *Corsochelys* because of its relatively large frontals and because the parietals are about as wide as they are long (Figure 4.7A, B). It differs from *C. haliniches* in that the frontals are more pointed anteriorly than in *C. haliniches*, the posterior margins of the frontals slope distinctly rearward from the midline instead of running directly at right angles from the midline suture as in *C. haliniches*, the frontals contribute less to the orbital rim than in *C. haliniches*, and the parietals meet posteriorly with only slight elongation rearward above the supraoccipital rather than forming an elongate cover above the supraoccipital as in *C. haliniches*. The humerus shaft is as wide and thick as in *Corsochelys*, but differs from *C. haliniches* in that the lateral tubercle forms a prominent knob that is not connected to the caput by a raised elongate ridge.

TYPE LOCALITY

S.C., Florence County; Burches Ferry, west bank of Pee Dee River, ca. 0.75 mile (1.21 km) east of County Road 52, ca. 0.08 mile (0.18 km)

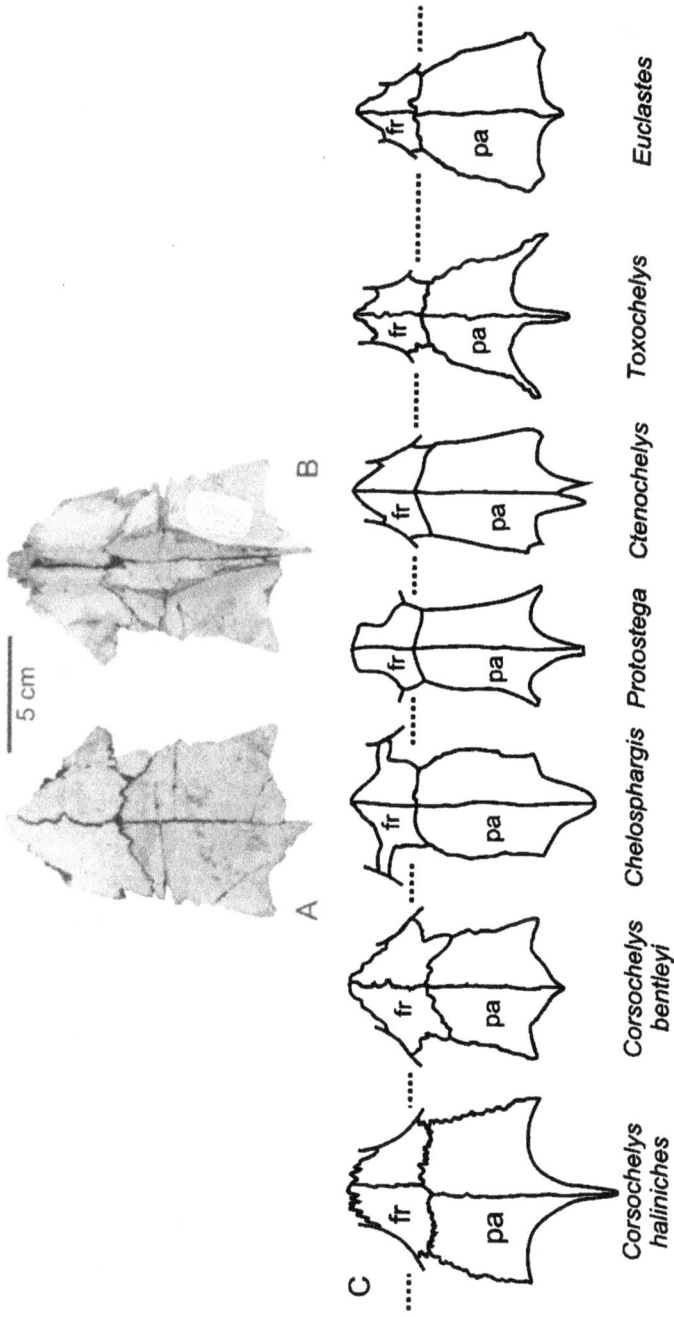

Figure 4.7 Dorsal (A) and ventral (B) views of SCSM 2005.25.3, holotype articulated frontals and parietals of *Corsochelys bentleyi* sp. nov. The frontals are unusually large and the parietals unusually small in *Corsochelys* compared to those in other Upper Cretaceous sea turtle taxa, as shown in the line drawings of the frontoparietal region of *Corsochelys* from Zangerl (1960) and that of five other well-known Upper Cretaceous North American sea turtles, *Chelosphargis* from Zangerl (1953) and *Protostega, Ctenochelys, Toxochelys,* and *Euclastes* from Hirayama (1994). All drawings are scaled to the same total length; dashed line shows the contact level between the frontals and the parietals in all taxa except *Corsochelys.*

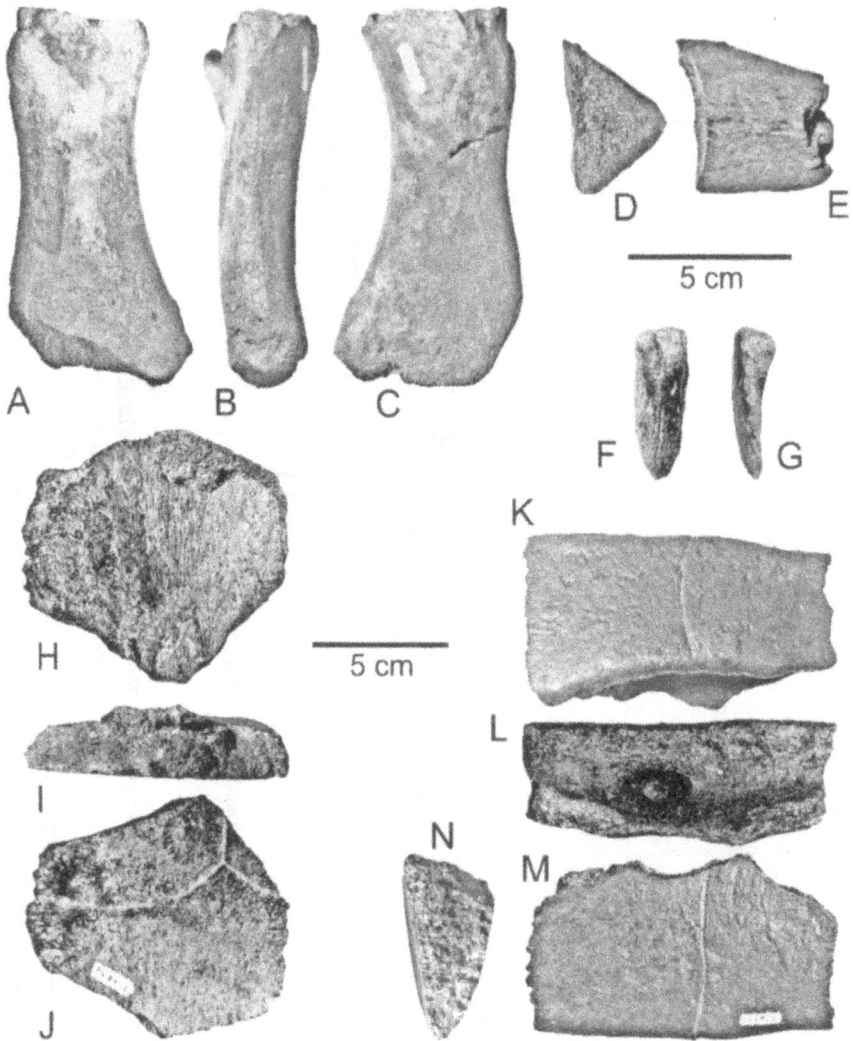

Figure 4.8. Limb and shell bones of *Corsochelys bentleyi* sp. nov. (A–C): ChM PV9132, right humerus in A, dorsal; B, posterior; and C, ventral views. (D, E): PV9147, medial peripheral fragment in anterior and external lateral views. (F, G): PV8934, ungual in ventral and lateral views. (H, I, J): PV8907, fragment of costal three or five near proximal end in ventral, proximal, and dorsal views. (K, L, M, N): PV8922, tenth right peripheral in ventral, internal, dorsal, and posterior views. Scale bar = 5 cm.

northwest of junction of Mill Branch and Pee Dee River. USGS Pamplico North 7.5′ quadrangle, 34° 03' 50″ N, 79° 31' 43″ W.

FORMATION AND AGE

Donoho Creek Formation, latest Middle–earliest Late Campanian; nannoplankton Zone CC22.

ETYMOLOGY

The specific name is a patronym honoring the collector, Curtis Bentley.

DESCRIPTION

Measurements of the holotype are as follows: Total preserved length of parietals and frontals: 127 millimeters. Maximum preserved width of parietals: 78.4 millimeters. Maximum preserved width at frontoparietal suture: 57 millimeters. Parietal length: 81.3 millimeters.

PARATYPES

ChM PV 8907, near proximal end of costal three or five; PV8922, tenth right peripheral; PV8934, ungual from manus; PV9132, shaft of right humerus; PV9147, fragment of medial peripheral. All paratype specimens are from site 1 (Stokes Quarry).

REFERRED MATERIAL

Site 1, Stokes Quarry—ChM PV7314, peripheral fragment; PV8910, nuchal fragment; PV8911, peripheral fragment; PV8915, second suprapygal.

DISCUSSION

Corsochelys haliniches originally was described by Zangerl (1960) as an advanced cheloniid sea turtle, but since then the genus has been recognized as an early dermochelyid (Hirayama 1992, 1994). The type material of that species is from the Mooreville Chalk of Alabama, which is upper Santonian to lower Campanian in age (Kiernan 2002). The material described here is very similar to that species, but significant minor differences are seen. The skull roof of the South Carolina specimen is far more similar to *Corsochelys* than it is to other North American Upper Cretaceous chelonioid skull roofs in that the frontals are exceptionally large and the parietals are more nearly equal in length and width (see Figure 4.7). There are, however, numerous minor differences that collectively indicate this material represents a distinct though closely related species. In the South Carolina specimen (see Figure 4.7A–B), the frontals are more pointed anteriorly than in *C. haliniches*, the posterior margins of the frontals slope distinctly rearward from the midline instead of running directly at right angles from the midline suture as in *C. haliniches*, the frontals contribute less to the orbital rim than in *C. haliniches*, and the parietals meet posteriorly with only slight elongation rearward above the supraoccipital rather than forming an elongate cover above the supraoccipital as in *C. haliniches*.

The humerus of the South Carolina species has a robustness and size similar to that of *C. haliniches* and also has the lateral tubercle in a position very similar to that of *C. haliniches*, but it differs in that the lateral tubercle is not connected to the caput by a high ridge (Figure 4.8A–C). All of these differences indicate that, although this material should be assigned to *Corsochelys*, it represents a different and somewhat younger species than *C. haliniches*. Other fragmentary material referable to this species includes a fragmentary, much reduced medial peripheral (PV9147; Figure 4.8D–E), an ungual from the manus flipper (PV8934; Figure 4.8F–G), a costal fragment from near (but not at) its proximal end (PV8907; Figure 4.8H–J), and a tenth right peripheral (PV8922; Figure 4.8K–N). None of this accessory material is complete enough to show any distinction from *C. haliniches*, but all of it is compatible with *Corsochelys*, based on its large size and extreme development of the costoperipheral fontanelles.

Family Toxochelyidae Baur, 1895; emended by Zangerl (1953)
Subfamily Toxochelyinae Zangerl, 1953
Toxochelys Cope, 1873
Toxochelys sp.
Figure 4.9D–I

MATERIAL

Site 1, Stokes Quarry—ChM PV8918, second or fourth neural.
Site 8, Burches Ferry—ChM PV7221, tenth left peripheral.

DISCUSSION

Five species of *Toxochelys* have been named over the years, but the most recent analysis of the genus by Nicholls (1988) concluded that *T. browni*, *T. weeksi*, and *T. barberi* are simply individual and age variants of *T. latiremis*. Thus only two species, *T. latiremis* and *T. moorevillensis*, are presently recognized in the Western Interior Seaway and Gulf Coast regions. No records of *Toxochelys* have been reported from the Atlantic Seaboard, so this material represents an eastward range extension of the genus. Both specimens from South Carolina are typical elements of *Toxochelys*, but neither the neural (PV8918; Figure 4.9D–F) nor the peripheral (PV7221; Figure 4.9G–I) has any characteristics that could distinguish between the two recognized species. Therefore, they can only be identified as *Toxochelys* sp. until more diagnostic material becomes available.

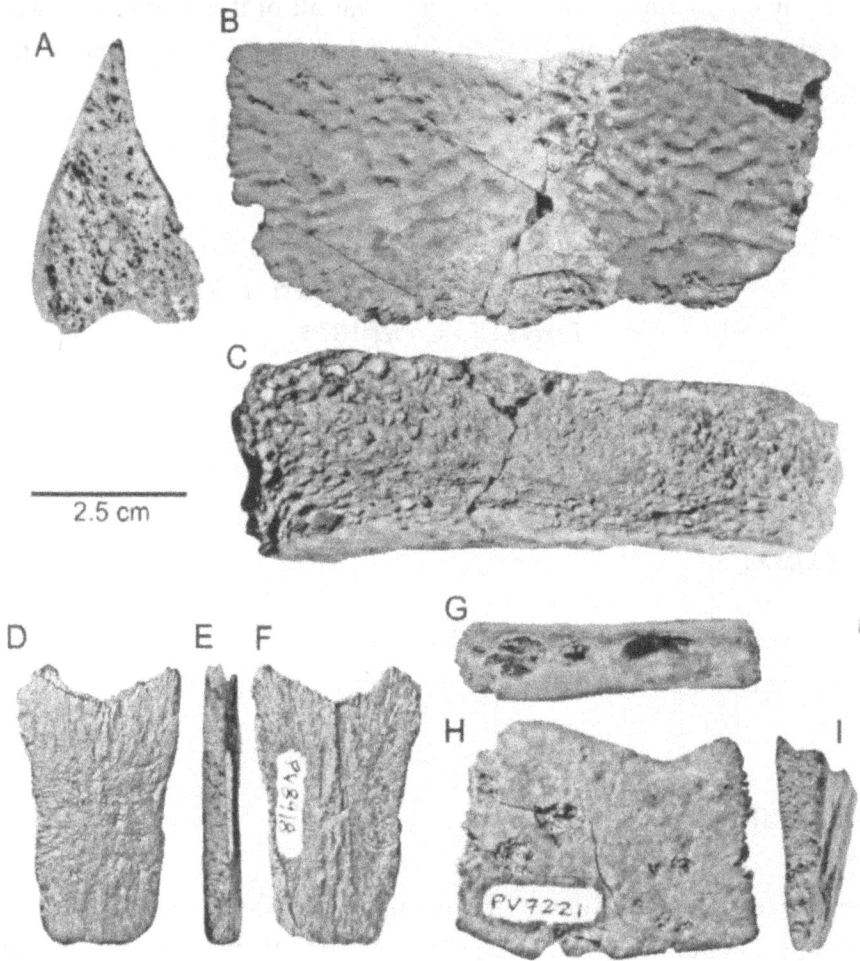

Figure 4.9. Carapace of toxochelyid turtles. (A, B, C): ChM PV8596, ninth right peripheral of *Peritresius ornatus* in anterior, dorsal, and internal views. (D, E, F): PV8918, neural of *Toxochelys* sp. In dorsal, right lateral, and ventral views. (G, H, I): PV7221, tenth left peripheral of *Toxochelys* sp. In internal, dorsal, and posterior views. Scale bar = 2.5 cm.

Subfamily LOPHOCHELYINAE Zangerl, 1953

Peritresius Cope, 1869c

Peritresius ornatus Leidy, 1856

Figure 4.9A–C

MATERIAL

Site 9, Pee Dee River near Allison Ferry Landing—ChM PV8596, peripheral.

DISCUSSION

The genus *Peritresius* was established by Cope (1869c) for a species earlier described by Leidy (1856) as *Chelone ornata*. The type specimen and other nineteenth-century occurrences were from the Maastrichtian Navesink greensands of Delaware and New Jersey (Baird 1964). Later, Hay (1908) reported an occurrence of *Peritresius* from the Maastrichtian Ripley Formation of Georgia (Baird 1964). Neither Hay (1908) nor Baird (1964) thought that the Georgia occurrence could be firmly assigned to the described species *P. ornatus*, but it is impossible to be sure whether the differences they observed are due to the Georgia material being a different species or whether the differences simply are due to age and individual variation. The present specimen extends the range of *P. ornatus* to the Maastrichtian Peedee Formation of South Carolina. Baird (1964), based on a relatively complete specimen from Sewell, New Jersey, determined that this species had prominent knobs on its hyoplastra and hypoplastra, a serrated carapace border, and a steep dorsal keel that runs along the midline of its neural elements. These traits are characteristic of lophochelyine toxochelyids (Zangerl 1953). Among lophochelyines, only *Peritresius* is known to have a trionychid-like sculpture on its carapace (Baird 1964). The peripheral described here (ChM PV8596; see Figure 4.9A–C) has this characteristic vermiculate sculpture pattern that radiates from the center of ossification of the bone and also a serrated border, so assignment to *P. ornatus* is certain. Based on the angle between the dorsal and ventral surfaces of this element, it is the ninth right.

Family PANCHELONIIDAE Joyce, Parham and Gauthier, 2004
Euclastes Cope, 1867

Zangerl (1953) synonymized fourteen species of five genera into a single taxon, *Osteopygis emarginatus*, pointing out its wide range of individual variation, especially regarding the shell. He further concluded that associated crania as well as other durophagous skull material belonged to *Osteopygis*. Fastovsky (1985) also assigned a durophagous skull from the Cretaceous of New Jersey to *Osteopygis*, and more recently Hirayama and Tong (2003) referred crania from the early Paleocene of Morocco to *Osteopygis emarginatus*. Parham (2005) regarded material previously referred to *Osteopygis* as representing a chimera and concluded that all of the durophagous sea turtle skull material referred to *Osteopygis* instead should be referred to *Euclastes* (Cope 1867), which was established originally on a headless carapace. This conclusion is supported by the great similarity between the carapace and plastron of *Osteopygis emarginatus* and the "macrobaenid" *Judithemys backmani* (Brinkman, Dinsmore, and Joyce 2010), and the mutual dissimilarity of both of these species with the carapace and plastron of *Erquelinnesia gosseleti* (Zangerl 1971), which has a durophagous skull very similar to those now assigned to *Euclastes wielandi*. Thus, skull material previously assigned to *Osteopygis emarginatus* is now reassigned to *Euclastes wielandi*. (For further discussion and a list of species now included in *Euclastes*, see Lynch and Parham 2003; Parham 2005.)

Euclastes wielandi (Hay, 1908)
Figures 4.10, 4.11

MATERIAL

Site 1, Stokes Quarry—ChM PV9001, a right quadrate, PV9002, left mandibular fragment; PV9003, right mandibular fragment; PV8703, right dentary; PV8919, PV8920, posterior peripherals, PV8908, PV8921, peripherals.
Site 2, Quinby— ChM PV6781, peripheral.

DISCUSSION

The snout of the skull of *Euclastes wielandi* is broadly rounded, has a wide flat mandible, a long mandibular symphysis that is shallow and about one third the length of the mandibular ramus, and a flat triturating surface (Parham 2005). The snout is comparable to that of recent *Caretta* (Hay 1908; Zangerl 1948; Hutchison and Weems 1998). Three dentaries (ChM PV8703, PV9002, PV9003) having these diagnostic features provide

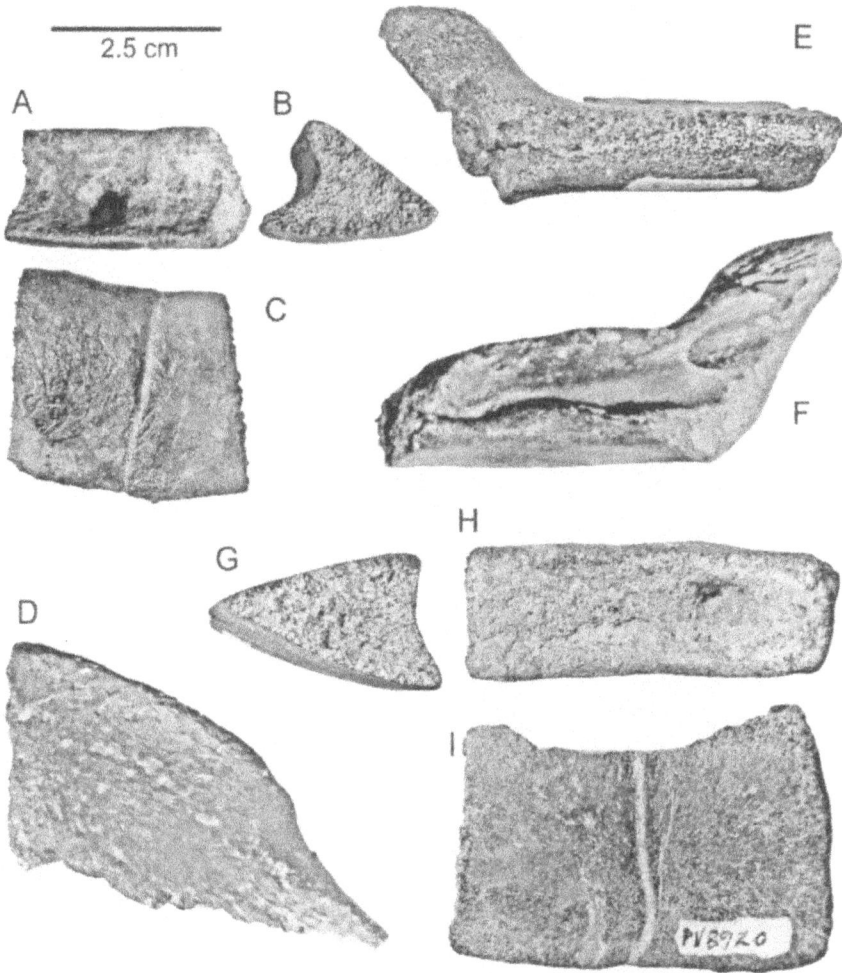

Figure 4.10. Skull and carapace of *Euclastes wielandi*. (A, B, C): ChM PV8919, posterior peripheral in internal, right–lateral, and dorsal views. (D, E, F): PV8703, right dentary in dorsal, external lateral, and internal lateral views, (G, H, I): PV8920, posterior peripheral in left lateral, internal, and dorsal views. Scale bar = 2.5 cm.

solid evidence of *Euclastes wielandi* in the assemblage from Stokes Quarry. A right lower jaw, PV8703 (see Figure 4.10D–F), is complete. It has a characteristic wide, flat, triturating area and a low lateral profile. Posteriorly, the dorsal surface is elevated between the foramen canalis alvoeolaris and the foramen canalis mentalis, separating the masseteric fossa and the sutural contact for the coronoid. The Meckelian groove medially reaches the posterior margin of the symphysis. The processus dentofacialis is

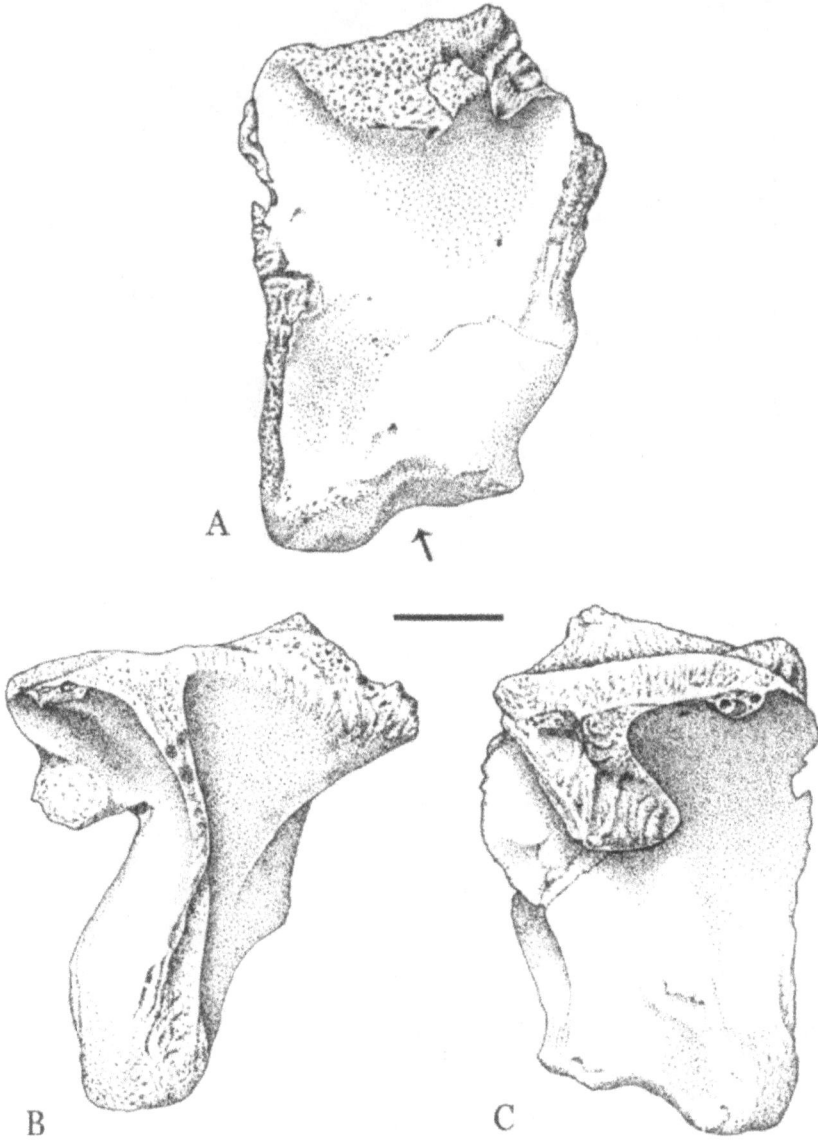

Figure 4.11. Right quadrate of *Euclastes wielandi*, ChM PV9001, in (A) anterior, (B) external, and (C) posterior views. Arrow points to the strongly stepped surface of the articular process. Scale bar = 1 cm.

indistinct. Two smaller fragmentary dentaries, PV9002 and PV9003, are right and left elements, respectively. The latter is missing its anterior portion from the symphysis forward, revealing the paths of the canalis mentalis and canalis alvoeolaris as described by Zangerl (1953) from X-ray analysis.

A right quadrate ChM PV9001 (see Figure 4.11) is the only skull element that can be positively identified. Notable is the strongly stepped surface of the articular process (see Figure 4.11B, arrow), which served to keep the jaws from becoming dislocated when they were applying a very strong biting force. This probably was a co-adaptation with the well-developed triturating surface of the lower jaws, which was an adaptation for crushing food. A few posterior peripheral elements also can be assigned to this taxon (PV8919; see Figure 4.10A–C; PV8920, see Figure 4.10G–I).

Order Sauropterygia Owen, 1960
Suborder Plesiosauria DeBlainville, 1935
Plesiosauria indet.
Figures 4.12, 4.13

MATERIAL

Site 1, Stokes Quarry—ChM PV8966, propodial; PV8967, girdle fragment.
Site 2, Quinby—ChM PV6768, vertebral centrum.
Site 3, Turbeville—ChM PV6827, vertebral centrum (not figured).
Site 8, Burches Ferry—SCSM 2005.25.4, vertebral centrum.
Site Number 9, Pee Dee River, two miles south of Allison Ferry Landing—ChM PV7181, three associated vertebral centra.

DISCUSSION

Three associated posterior cervical vertebral centra (ChM PV7181), found in place in the Donoho Creek Formation, evidently belong to a single individual but are not sequential in their relationships to one another. All three vertebrae show little indication of taphonomic damage. One

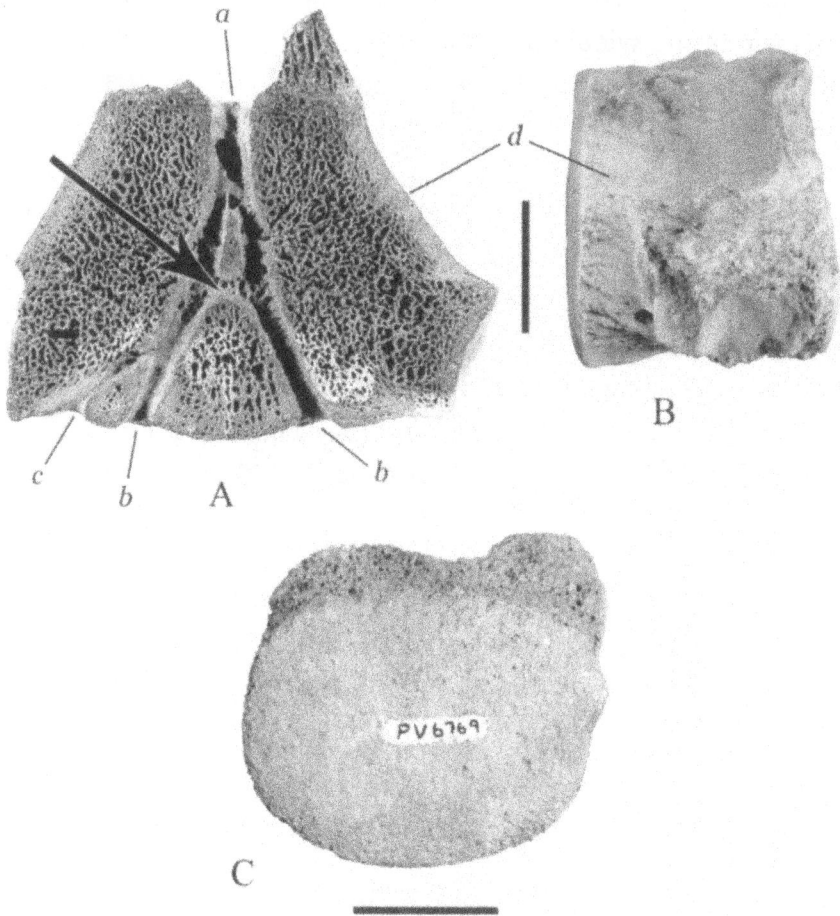

Figure 4.12 Vertebrae, Plesiosauria, indeterminate. (A): ChM PV7181, caudal vertebra, sectioned view: (a): nutrient foramen, dorsal; (b): nutrient foramen ventral; (c): secondary nutrient foramen, ventral; (d): cortical bone surface. Arrow points to nutrient canal link. (B): associated vertebra (PV7181), left lateral view. (C): PV6769, vertebral centrum with ablated cortex. Scale bars = 3 cm.

specimen (Figure 4.12A) has lost most of its neural arch and left rib facet as a result of fracturing and separation after fossilization. Its centrum has been transversely sectioned at about midlength to illustrate the division of the main nutrient canal system, which originates in the cortical bone as a large median foramen located on the floor of the neural canal. Within the complex medullary portion of the vertebral body, the main nutrient

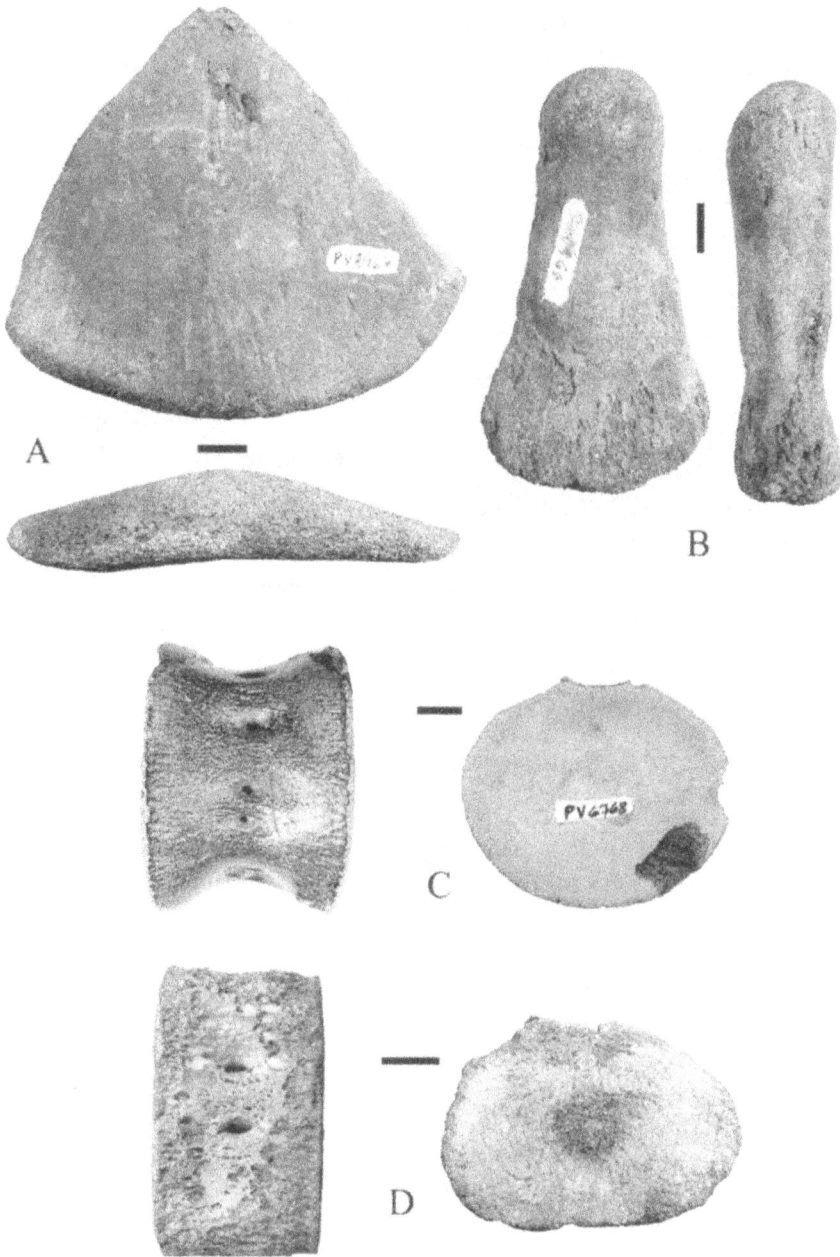

Figure 4.13 Plesiosauria, indeterminate. (A): Probable coracoid (ChM PV8967), dorsal and medial views. (B): Indeterminate podial (PV8966), dorsal (left) and lateral (right) views. (C): Vertebral centrum (PV 6768), ventral and end views. (D) Vertebral centrum (SCSM 2005.25.4), ventral and end views. Scale bars = 1 cm.

canal divides into two slightly smaller canals that are cross-linked (Figure 4.12A, arrow) and become widely separated before emerging through prominent foramina on either side of the midline on the ventral surface of the centrum. In this specimen, auxiliary branches exit through smaller foramina laterally located on the ventral surface. Judging from the multiple foramina on the ventral and lateral surfaces of other specimens, the auxiliary system was variable, and probably is indicative of the position of the vertebra within the vertebral column (Welles 1943). Cortical bone is indicated in Figure 4.12A and B. Another plesiosaur bone, PV6769 (Figure 4.12C), reworked from the Bladen Formation, differs from PV7181 in having lost its entire cortical surface. These two specimens represent the largest plesiosaurs in the assemblage, and may belong to the same taxon occurring at two localities in Florence County.

Among other plesiosaur remains available are several undifferentiated elements from Campanian and Maastrichtian deposits showing varying degrees of completeness. A fragment of a flat, plate-like bone, probably a coracoid (ChM PV8967; Figure 4.13A), retains much of its surface detail with a central thickening of the midline suture, not characteristic of the pectoral girdle (Welles 1962). A second plesiosaur bone, PV8966 (Figure 4.13B), is an eroded propodial, based on its general form (Welles 1943, 1962; Romer 1956). Its small size suggests a young individual. ChM PV6768 and SCSM 2005.25.4 (Figure 4.13C, D) are vertebral centra, which offer little diagnostic information other than to illustrate the variation in size as well as the location of the foramina for the nutrient canal system discussed previously.

Order SQUAMATA Oppel, 1811
Family TEIIDAE Gray, 1827
Teiidae, genus indet.
Figure 4.14

MATERIAL

Site 1, Stokes Quarry—ChM PV8980, vertebra.

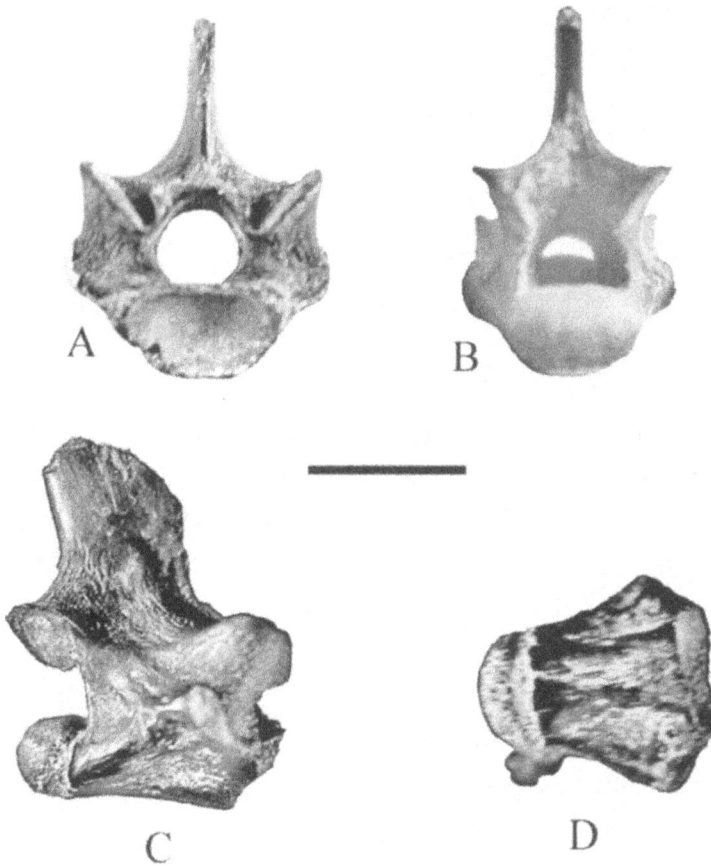

Figure 4.14 Trunk vertebra of teiid lizard (ChM PV8980). (A): Anterior view. (B): Posterior view. (C): Right lateral view. (D): Ventral view. Scale bar = 5 mm.

DISCUSSION

A complete dorsal vertebra, PV8980 (Figure 4.14A–D), of a small lizard is characterized by a procoelous centrum and prezygapophysial facets that are medially upturned to the extent of having prezygapophysial and zygosphenic facets nearly opposite each other. This condition is typical of the family Teiidae (Gilmore 1942; Hoffstetter and Gasc 1969). Unlike snakes, the upper lips of the zygosphenic facets are deeply notched. The centrum is dorsoventrally compressed, and ventrally a pair of low rounded ridges run the length of the centrum.

Family MOSASAURIDAE Gervais, 1853
Tylosaurus Marsh, 1872
***Tylosaurus* sp**
Figures 4.15, 4.16A–D

MATERIAL

Site 1, Stokes Quarry—ChM, PV8964, tooth.
Site 7, Diamondhead Loop Road—PV7680, tooth.
Site 8, Burches Ferry—SCSM 2005.25.18, quadrate; SCSM 78.16.6, ChM PV6779, teeth.

DISCUSSION

A left quadrate (SCSM 2005.25.18; Figure 4.15A–D) is the only mosasaur skull bone recovered. It is mostly complete, missing only the end of the suprastapedial process, infrastapedial process, marginal details of the anterior rim, and the articulating surface of the mandibular condyle. These missing pieces do not allow for definitive comparisons with described *Tylosaurus*, however, this bone closely resembles the specimen figured in Bell (1997, fig. 7B), as well as a quadrate figured by Russell (1967, fig. 94B), both identified as *Tylosaurus nepaeolicus*. Proportionally, it is taller and less robust than the quadrates of *Platecarpus*, *Clidastes*, and *Mosasaurus*. The stapedial pit, anterior to and slightly superior to the top of the stapedial notch on the medial surface, has a rectangular rather than elliptical shape as in *Clidastes*, *Mosasaurus*, and *Platecarpus*, or the circular shape in *Prognathodon* (Russell 1967; Bell 1997). The distal end of the external surface of the suprastapedial process, which is only moderately large, unlike that of *Platecarpus*, projects anteroventrally, and would have descended in a circle back toward the ventral portion of the shaft. The missing infrastapedial process was evidently small and may not have joined the suprastapedial process to enclose a long oblong stapedial notch

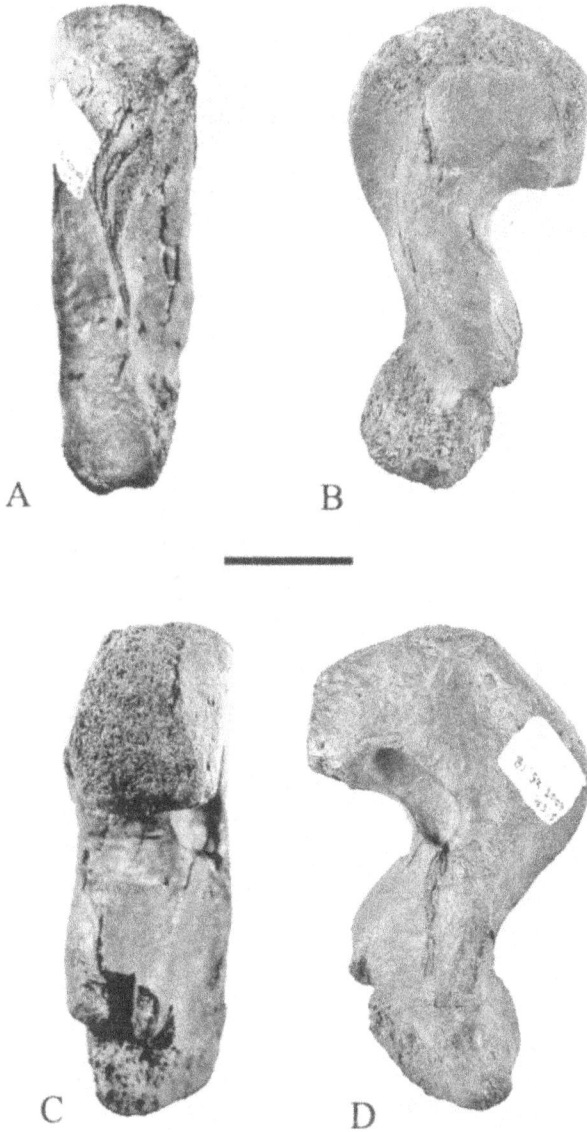

Figure 4.15 Left quadrate, *Tylosaurus* sp. (SCSM 2005.25.18). (A): Anterior view. (B): Lateral view. (C): Posterior view. (D): Medial view. Scale bar = 2 cm.

as in *Platecarpus planifrons* (Konishi and Caldwell 2007). The stapedial notch was evidently open posteriorly.

With regard to the typically isolated mosasaur material found in the eastern United States, Baird (1986) pointed out the strong possibilities

Figure 4.16 Teeth of *Tylosaurus* sp. (A–D) and *Prognathodon?* sp. (E). (A): SCSM 78.16.6, (B): ChM PV6779, (C): PV8964, (D): PV7680, tooth crowns of *Tylosaurus* sp. (E): SCSM 84.176.1, tooth crown of *Prognathodon?* sp., anterior and lateral views. Scale bars = 1 cm.

for error with positive identification of most isolated mosasaur teeth. With this caveat in mind, we tentatively identify only a few mosasaur teeth here. Carinate, typically recurved mosasaur teeth (SCSM 78.16.6, ChM PV6779, PV7680, PV8964; Figure 4.16A–D) are here referred to as belonging to *Tylosaurus* sp. because they feature strong vertical fluting of the external enamel surfaces. These flutes occur on both lateral surfaces of the crown, with six to eight concave facets between carinae. The surfaces are also finely striated within the flutes.

Prognathodon Dollo, 1889
Prognathodon? sp.
Figure 4.16E

MATERIAL

Site 13, Myrtle Beach—SCSM 84.176.1, tooth.

DISCUSSION

The isolated tooth SCSM 84.176.1 (Figure 4.16E), of probable Maastrichthian age (?Peedee Formation) from the vicinity of Myrtle Beach in Horry County (Site 13), is a large posterior marginal tooth crown with prominent but rounded anterior and posterior carinae. The lingual and buccal surfaces are subequal in shape and the surface is covered with smooth thin enamel. It is identified here as *Prognathodon*? because of its subrounded base dimensions, rounded carinae, and enamel surface lacking evident surface fluting, which suggests crushing rather than shearing bite forces, as the jaws and teeth of *Prognathodon* indicate in well-preserved individuals from other regions.

Mosasauridae indet.
Figures 4.17, 4.18

77

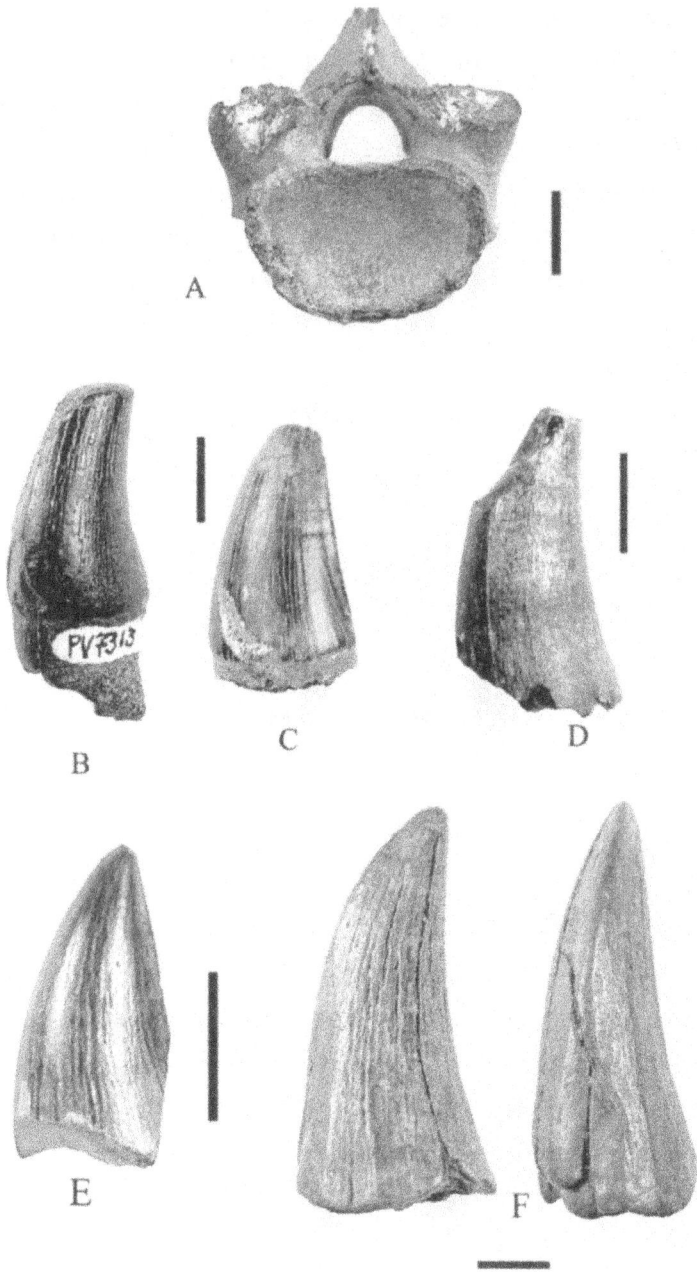

Figure 4.17 Vertebra and teeth of indeterminate mosasaurs. (A): ChM PV8310, cervical vertebra, anterior view. (B): PV7313, (C): PV6828, (D): PV8591, (E): PV6780, teeth in lateral view. (F): SCSM 94.107.1, tooth, buccal and lingual views. Scale bars = 1 cm.

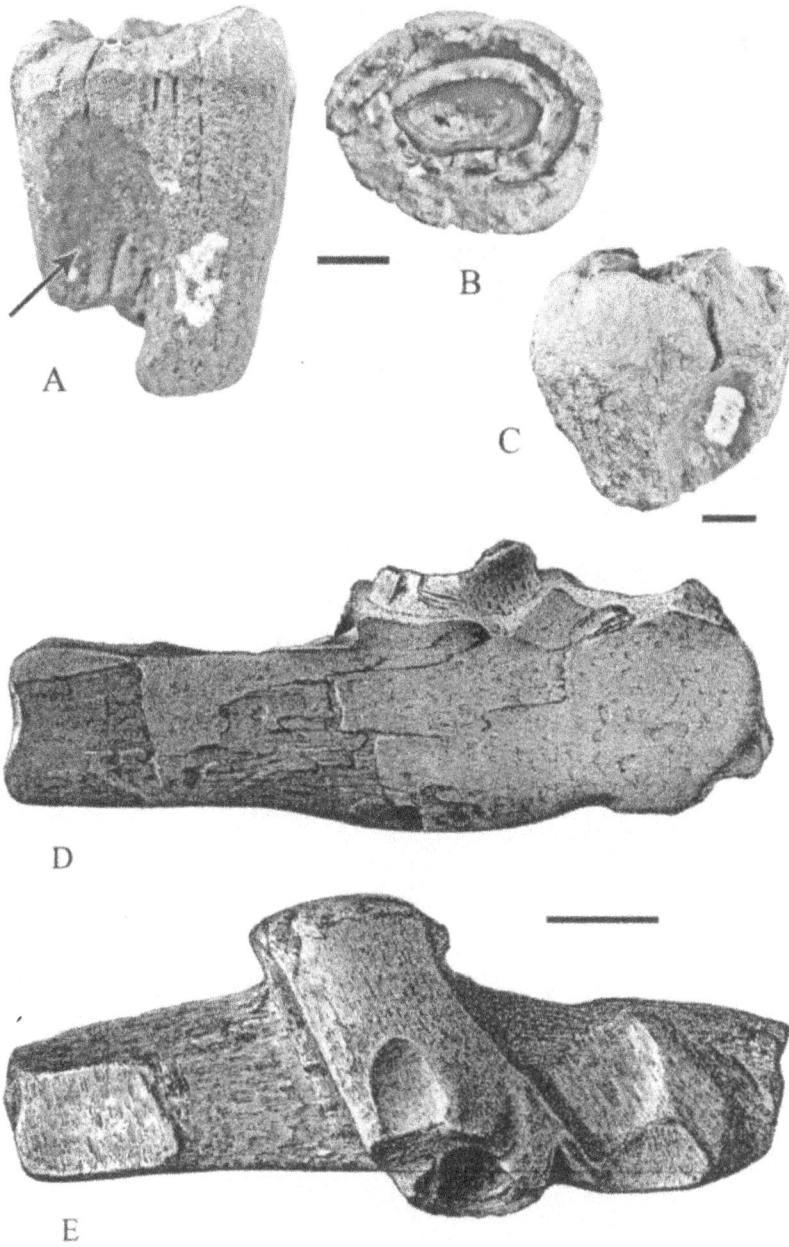

Figure 4.18 Teeth referred to indeterminate mosasaurs and the lost holotype mandible of *Mosasaurus caroliniensis* (Gibbes, 1851, pl. 2). (**A**): ChM PV7332, tooth base with resorption cavity (*arrow*) in lingual view and (**B**): occlusal view. (**C**): PV8696, tooth base with resorption cavity, lingual view. Holotype mandible of *Mosasaurus caroliniensis* (Gibbes 1851) in (**D**): lateral view, and (**E**): inverted lingual view. Scale bars: A–C = 1 cm; D, E = 25 mm.

MATERIAL

Site 1, Stokes Quarry—ChM PV7313, PV8310, cervical vertebra; PV8696, PV8965, vertebral centra; PV7332, PV7338, PV7339, PV7340, PV8592, PV8964, PV9118, teeth.
Site 3, Turbeville—PV6828, tooth.
Site 7, Diamondhead Loop Road—ChM PV7312, tooth (not figured here).
Site 8, Burches Ferry—PV6780, tooth.
Site 12, Waccamaw River—SCSM 94.107.1, tooth (not figured here).
Site 13, Myrtle Beach—SCSM 84.176.1, tooth (not figured here).

DISCUSSION

Among the tooth and vertebral specimens that are clearly from mosasaurs but not identifiable to the genus, ChM PV7332 and PV8696 (Figure 4.18A–C) illustrate the typical implantation of marginal teeth and an early stage in the development of the successional tooth process in mosasaurs. In each specimen, only the tooth base and base of the tooth crown has been preserved. As pointed out by Russell (1967), the grain of the tooth base characteristically parallels the longitudinal axis of the tooth base. This is clearly visible in each specimen, as is the thick layer of dentine on each tooth. The base of each tooth also exhibits an invasion (resorption) cavity created by a developing replacement tooth on its posteromedial/distolingual side, near the alveolar margin. This condition is part of the tooth replacement process, which occurs in all mosasaurs with broadly conical, recurved teeth. Despite their large size and elliptical shape, without crowns for comparison, these specimens are not informative enough for reliable taxonomic allocation below the family level. This provision applies also to the lost holotype of Gibbes's (1851) *Mosasaurus caroliniensis* (Figure 4.18D, E), which Williston (1897) referred to *Tylosaurus,* perhaps because of its size or the elliptical shape of the alveoli.

CROCODYLIFORMES Hay, 1930
MESOEUCROCODYLIA Whetstone and Whybrow, 1983
EUSUCHIA (=CROCODYLIA Huxley, 1875)

Four eusuchian crocodyliform taxa (here referred to for clarity as "crocodilians") have been identified from specimens at eight of the fourteen sites investigated during this study. Crocodilians, representing a range of habitats, contribute the second-most abundant reptile remains in the collective assemblage, as might be expected in the marginal marine settings of the region.

Deinosuchus Holland, 1909
Deinosuchus rugosus (Emmons, 1856)
Figure 4.19A–C

MATERIAL

Site 1, Stokes Quarry—ChM PV8962, tooth.
Site 2, Quinby—ChM PV6774, tooth.
Site 11, Kingstree—ChM PV7304, tooth.

DISCUSSION

Three teeth, ChM PV6774, PV7304, PV8962 (Figure 4.19A–C), have the diagnostic features of *Deinosuchus rugosus* (Emmons 1856). They are relatively large (basal diameters > 2 cm) and robust, with thick, infolded enamel and dentin, and bluntly rounded tips on the crowns. A notable characteristic of the teeth is their massive cross-sections, with multiple replacement cusps stacked internally, leaving very little interior space.

Osteoderms, as well as teeth, are among the most diagnostic meristic features of *Deinosuchus* (Schwimmer 2002). Yet, among the numerous fragmentary osteoderms in the assemblage, none show the features described for *Deinosuchus* (or *Phobosuchus*) by Holland (1909), Colbert and Bird (1954), and Schwimmer (2002), as being extraordinarily heavy, thick, often irregular in shape, and with large, deep pits.

Borealosuchus Brochu, 1997
Borealosuchus sp. indet.
Figure 4.19 D–H

81

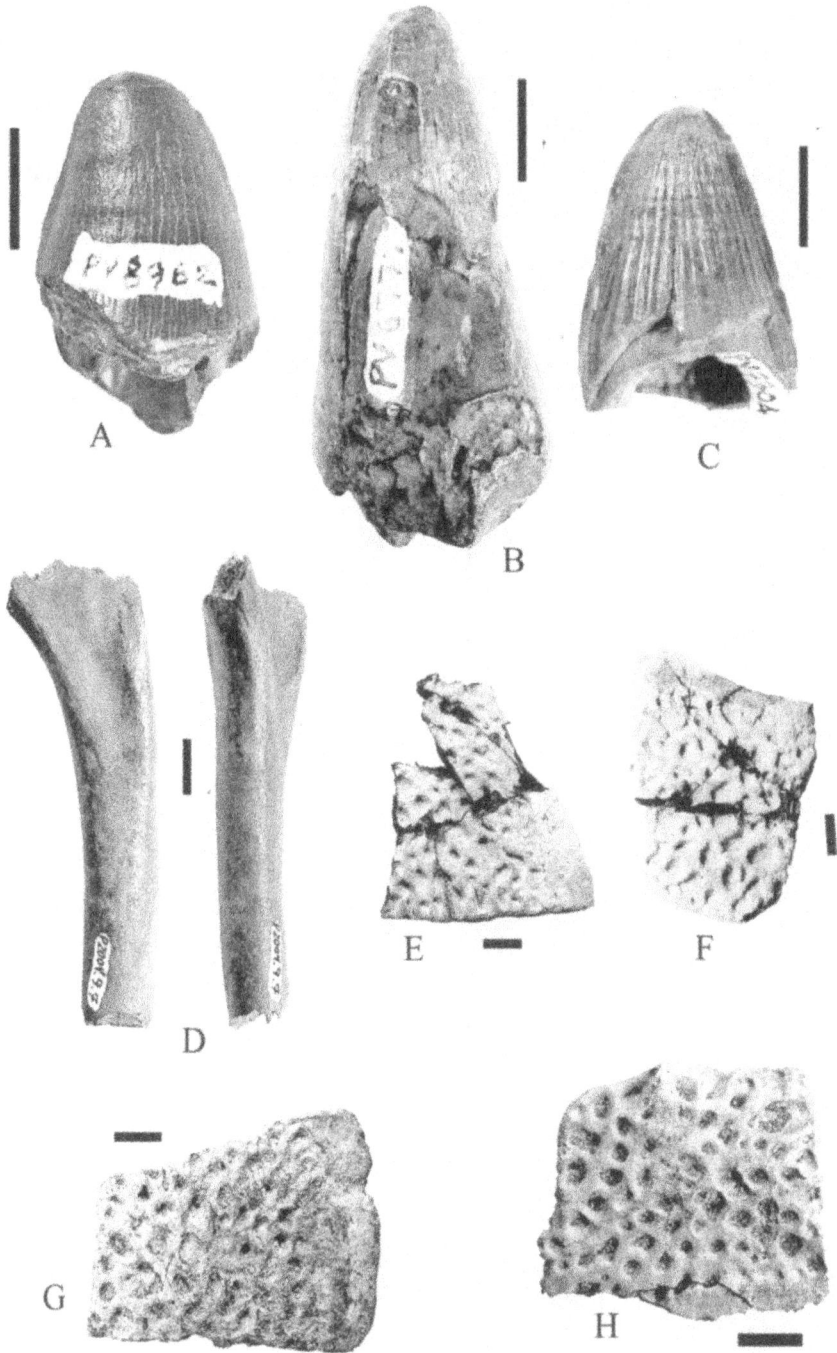

Figure 4.19 Teeth of *Deinosuchus rugosus* (A–C) and humerus and osteoderms of *Borealosuchus* sp. (D–H). (A): ChM 8962, (B): PV6774, (C): PV7304, tooth crowns of *Deinosuchus rugosus*. (D): SMM P2004.9.7, proximal half of left humerus, *Borealosuchus* sp. in ventral and posterior views. (E): SCSM 2005.25.7, (F): 2005.25.8, (G): ChM PV6770, (H): PV8843, osteoderms, *Borealosuchus* sp., external view. Scale bars A–C = 3 cm; D–H = 1 cm.

MATERIAL

Site 1, Stokes Quarry—SMM P2004.9.7, proximal half of left humerus.
Site 2, Quinby—PV6770, osteoderm fragment.
Site 7, Diamondhead Loop Road—ChM PV8843, osteoderm fragment.
Site 8, Burches Ferry—SCSM 2005.25.7, 2005.25.8, osteoderm fragment.

DISCUSSION

"*Leidyosuchus* cf. *L. formidabilis*" (= *Borealosuchus formidabilis* [Erickson 1976, Brochu 1997]) was reported from the Black Creek deposits of North Carolina by Baird and Horner (1979) as an "ordinary sized" crocodilian associated with *Deinosuchus*. Hence, it is not surprising that *Borealosuchus* appears in the present assemblage.

The proximal half of a left humerus (SMM P2004.9.7, Figure 4.19D) is here referred to *Borealosuchus* because of its gracile morphology and the pronounced anterior curve of the proximal end of the shaft, with its characteristic deltopectoral crest. Among other crocodilians identified in the assemblage, for example, *Deinosuchus*, the end of the humeral shaft is relatively straight.

Several osteoderm fragments (ChM PV6770, PV8843, SCSM 2005.25.7, SCSM 2005.25.8 (Figure 4.19E–H) are thin and flat, some slightly raised at the center, with generally small circular and subcircular pits, and with a thin, smooth imbricating margin. These are also characteristic of *Borealosuchus*.

Bottosaurus Agassiz, 1849
Bottosaurus sp. indet.
Figure 4.20A–D

MATERIAL

Site 1, Stokes Quarry—ChM PV7379, PV8896, PV9152, tooth crowns.
Site 11, Kingstree—SMM P2006.1.14, tooth.

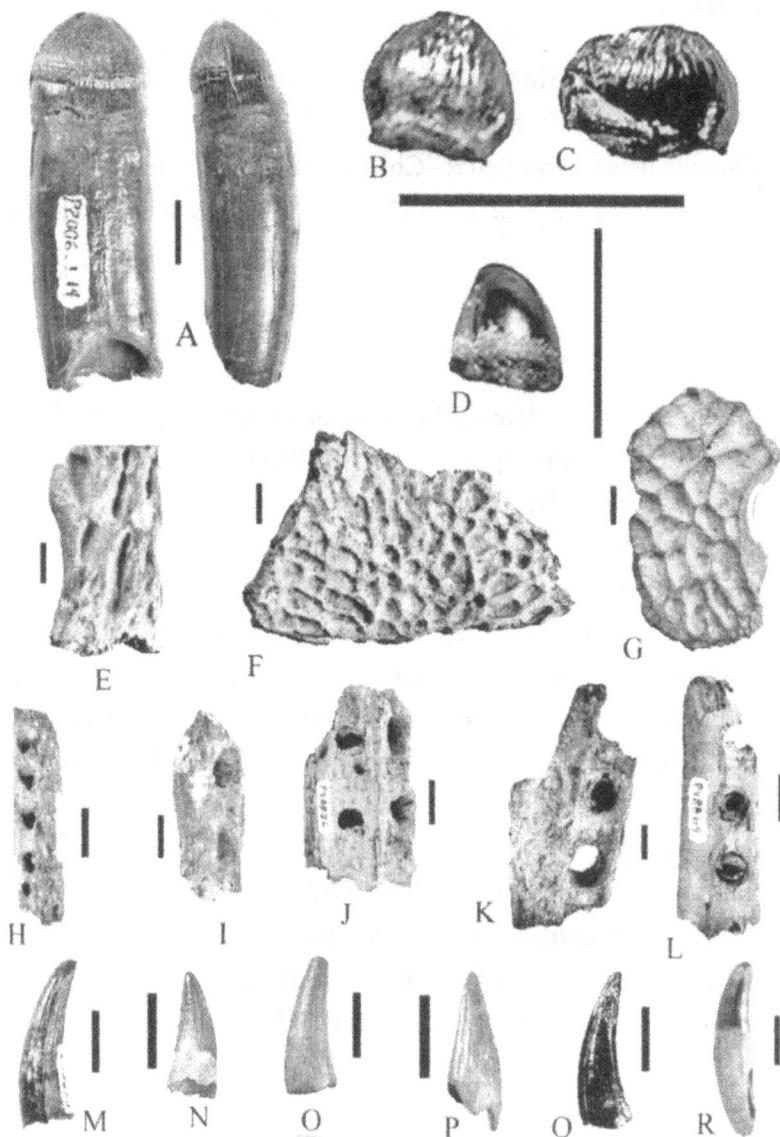

Figure 4.20. Teeth of *Bottosaurus* sp. indeterminate (A–D) and bones and teeth of indeterminate gavialoid crocodilians (E–R). (A): SMM P2006.1.14, tooth in lingual and posterior views. (B): ChM PV7379. (C): PV9152. (D): PV8896, tooth crowns of *Bottosaurus* sp. indeterminate. (E): ChM PV6776, squamosal fragment, dorsal view. (F): PV8706, frontal less crista crania frontalis, dorsal view. (G) PV8892, parietal fragment, dorsal view. (H): ChM PV8874, maxillary fragment, occlusal view. (I): PV8875, maxillary fragment, occlusal view. (J): PV8876, maxillary fragment, oblique view. (K): SM M P2004.9.11, right maxillary fragment, occlusal view. (L): PV8869, dentary fragment, occlusal view. (M): PV8868, tooth, labial view. (N): PV5873, tooth, labial view. (O): PV6772, tooth, lingual view. (P): PV6773, tooth, labial view. (Q): PV8842, tooth, labial view. (R): PV8859, tooth, anterior view, Gavialoidea indeterminate. Scale bars: A = 1 cm; B = 5 mm; C–R = 1 cm.

84

DISCUSSION

A third crocodilian, *Bottosaurus*, is recognized at two locations. At Kingstree (Site 11), it occurs in the Late Cretaceous Donoho Creek sediments as well as in the overlying late Paleocene deposits of the Williamsburg Formation (Erickson 1998). A complete tooth (SMM P2006.1.14; Figure 4.20A) from the Donoho Creek sediments at Kingstree, with root intact and showing a resorption cavity (notch) at its base, closely matches the teeth of *Bottosaurus harlani* (NJSM 11265) from the late Maastrichtian New Egypt Formation of New Jersey (D. Baird, personal communication, to Bruce R. Erickson). The crown is low with a nearly circular cross-section. Its surface has fine vertical striae and the base is slightly constricted. Anterior and posterior carinae are present and the apex is bluntly pointed (Mook 1925). This tooth could easily be mistaken for that of *Brachychampsa montana* (Gilmore 1911).

Stokes Quarry (Site 1) yielded two smaller tooth crowns (Figure 4.20B, C), ChM PV7379 (crown height 2.5 mm) and PV9152 (crown height 6.0 mm) that have incipient carinae near their apices. These small teeth with rugose enamel surfaces and a constricted base are similar to a tooth crown height (4.5 mm) described by Miller (1968) as *Bottosaurus?* sp. The absence of carinae in these small teeth may be a reflection of age or size and they are tentatively referred to *Bottosaurus*. A third uneroded tooth (PV8896; Figure 4.20D) has strong carinae that run from the apex to the base of the crown.

Gavialoidea Hay, 1930
Gavialoidea indet.
Figures 4.20, 4.21A–B

MATERIAL

Site 1, Stokes Quarry—ChM PV8706, frontal fragment; PV8874, PV8875, PV8876, maxillary fragments; PV8892, parietal fragment, PV8869, PV9139, left dentary fragments; SMM P2004.9.11, right maxillary fragment, ChM PV8832, PV8842, PV8859, PV8860, PV8868, PV8873, PV8879, PV8887, teeth.

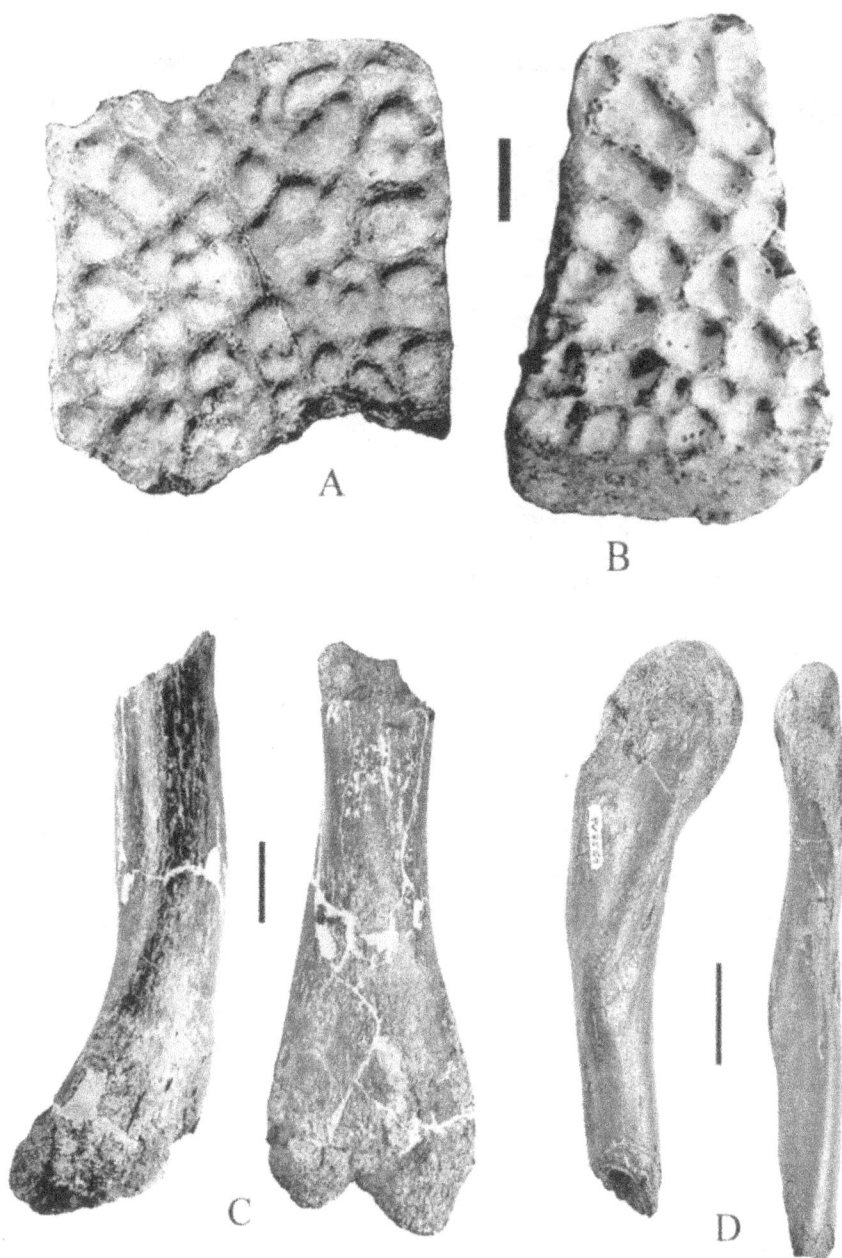

Figure 4.21. Osteoderms of Gavialoidea indeterminate (A–B) and femora of indeterminate eusuchian crocodilians (C–D). (A): ChM PV6829 and (B) PV6849, gavialoid osteoderms. (C) SCSM 2005.11.2, distal portion of eusuchian femur. (D): ChM PV7559, proximal portion of femur. Scale bars: **A, B** = 1 cm; **C, D** = 3 cm.

Site 2, Quinby—PV5873, PV5878, PV5879, PV5880, PV6772, PV6773, teeth; PV6771, femoral fragment; PV6829, osteoderm fragment.
Site 3, Turbeville—ChM PV6776, squamosal fragment; PV6849, PV6850, PV6851, osteoderm fragments.

DISCUSSION

The presence of one or more gavialoid crocodilian genera is indicted by cranial fragments, isolated teeth, and osteoderms; however, the material is too fragmentary to provide definitive evidence to separate two similar Late Cretaceous longirostrine genera: *Eothoracosaurus* (Brochu 2004) and *Thoracosaurus* (Leidy 1852). Among the diagnostic characters separating these genera (Brochu 2004) are details of the width and anterior (rostral) projection of the frontals, presence of conjoined third and fourth dentary teeth in *Eothoracosaurus*, and the relative width of the skull table, none of which are documentable from the preserved material. ChM PV6776 (Figure 4.20E) is a squamosal fragment indicating a flat cranial table and large supratemporal fenestra. PV8706 (Figure 4.20F) is a frontal lacking the crista crania frontalis. This frontal is relatively broad and more closely resembles the holotype of *Thoracosaurus mississippiensis* (Brochu 2004; Carpenter 1983); however, lacking the anterior projection, this is inconclusive evidence. ChM PV8892 (Figure 4.20G), a fragment of the right side of the parietal, including part of the supratemporal fenestra, has a slightly elevated rim. The skull table is flat with large, closely spaced pits.

Maxillary fragments PV8874, PV8875, PV8876, and SMM P2004.9.11 (Figure 4.20H–K) each preserve several alveoli, some with broken teeth in place. The widely spaced alveoli with intervening occlusion pits for reception of dentary teeth are features of gavialoid eusuchians (Mook 1925; Carpenter 1983; Erickson 1998). By contrast, the alveoli of contemporary brevirostrine crocodilians *Deinosuchus* and *Borealosuchus* are closer together, and occlusion pits for lower jaw teeth are located mostly medial to the alveoli.

A fragment of the left dentary (PV8869; Figure 4.20L) does not have notably elevated alveolar rims as in the type specimen of *Eothoracosaurus* (Brochu 2004), but this may be an artifact of preservation.

Isolated teeth of most crocodilians are rarely diagnostic to genus, aside from distinctive morphologies as in *Deinosuchus* and *Bottosaurus*. Teeth from four of the localities in the present study are identified only as Gavialoidea indet. based on their slender, curved overall shape. All included here have lateral carinae ranging from sharp to indistinct. Several show distinct fluting (e.g., PV 8868, 8842, 6772, 6773; Figure 4.20M–P), whereas others have surfaces that are relatively smooth to finely striated (e.g., PV8859, PV5873; Figure 4.20Q–R). Species attributed to *Thoraco-saurus* have been described showing both morphologies (e.g., Troxell 1925), whereas the type specimen of *Eothoracosaurus* has only maxillary teeth preserved, but with poorly preserved enamel surfaces, thus leaving no clear distinction among the genera. Likewise, two osteoderm frag-ments, PV6829 and PV6849 (Figure 4.21A–B), are nondescript: each has a beveled, imbricating edge and closely spaced pits, which could be attributed to either genus.

Eusuchia indet.
Figures 4.21C–D, 4.22–4.24

MATERIAL

Site 1, Stokes Quarry—ChM PV8858, complete articular with apparent bite marks (see Figure 5.3F, p. 116); PV8839, PV8878, articulars; PV9151, skull and mandible fragment, PV8593, cranial fragment; PV8866, jugal; PV7365, PV7561, PV8312, PV8676, PV8836, PV8837, PV8840, PV8841, PV8863, PV8864, PV8865, PV8882, PV8883, PV8884, PV8897, PV9097, incomplete vertebrae; PV8894, coracoid (hatchling); PV8895, right humerus (hatchling); PV8854, PV8872, cervical ribs; PV9138, dorsal rib; PV8830, presacral diapophysis; SMM P84.12.26, anterior zygapophysis; SCSM 2005.11.2, distal portion of femur.
Site 2, Quinby—ChM PV6826, skull fragment.
Site 3, Turbeville—PV6849, PV6850, PV6851, osteoderm fragments (not figured here).
Site 5, Black Creek—ChM PV8848, parietal.

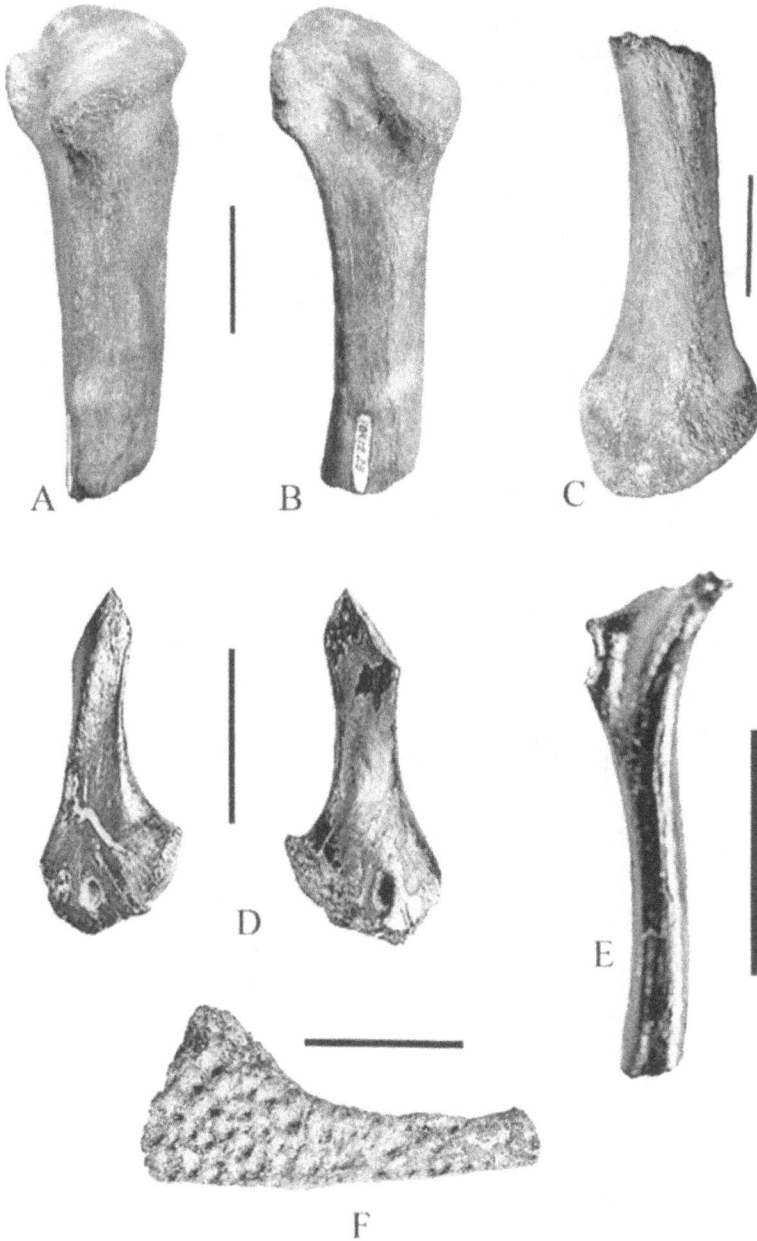

Figure 4.22. Limb and pectoral bones of indeterminate eusuchians. (A–C): SMM P84.12.28, distal half of left tibia. (D): PV8894, right coracoid, hatchling individual. (E): PV8895, right humerus, hatchling individual. (F): PV9151, left part of frontal, juvenile individual. Scale bars: A–C = 3 cm; D–F =1 cm.

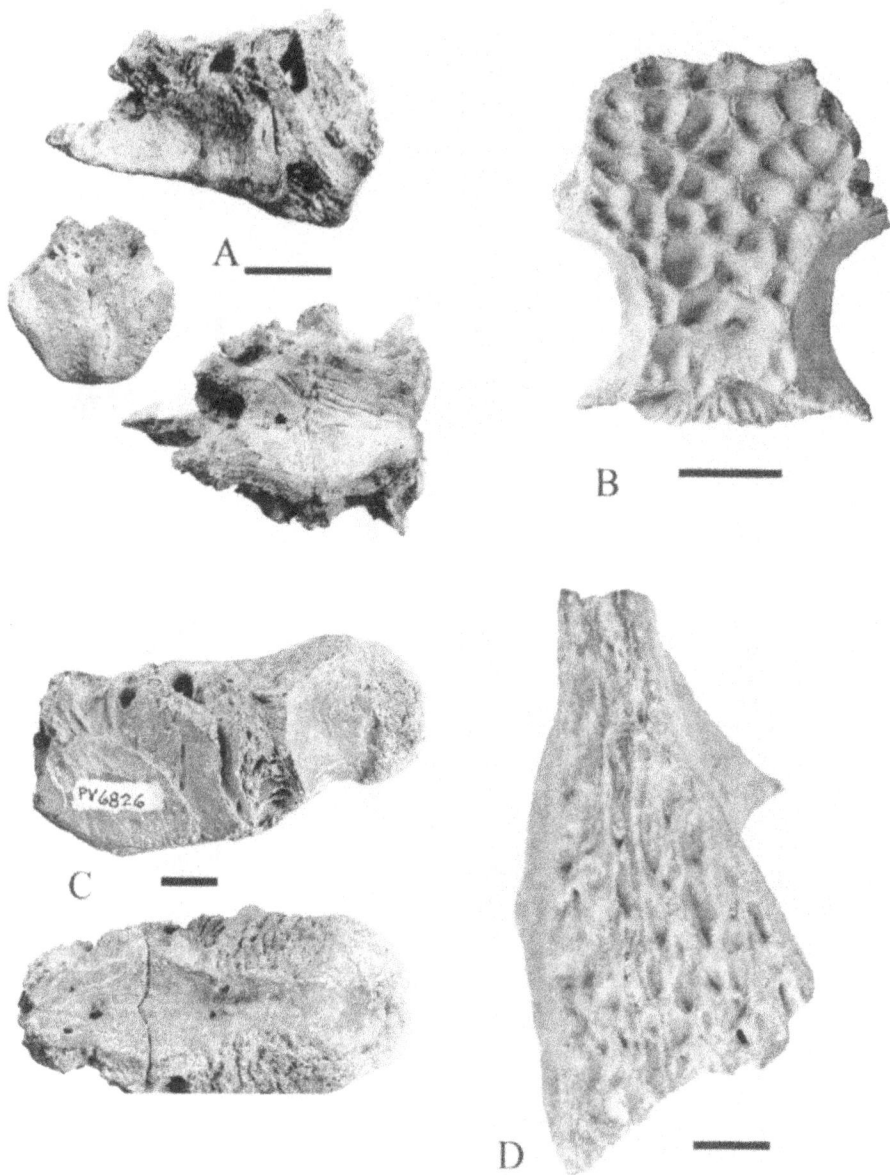

Figure 4.23. Cranial bones of indeterminate eusuchian crocodilians. (A): ChM PV8593, partial basisphenoid and basioccipital. (B): PV8848, parietal. (C): PV6826, basisphenoid/basioccipital complex. (D): PV8866, jugal. Scale bars = 1 cm.

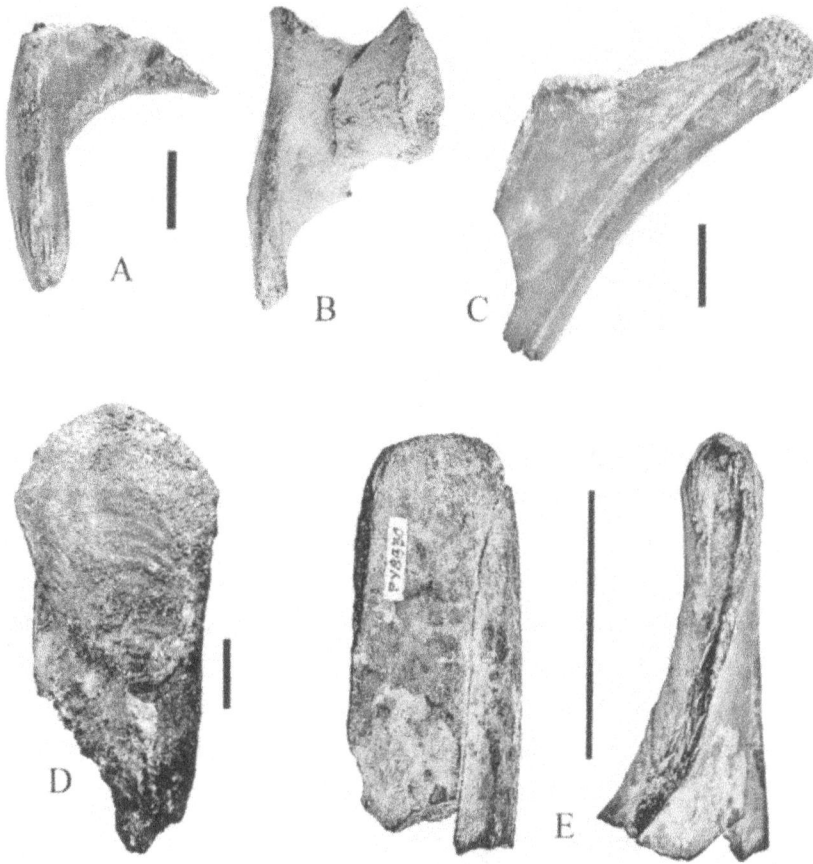

Figure 4.24 Postcranial bones of indeterminate eusuchian crocodilians. (A): ChM PV8854. (B): PV8872, cervical ribs. (C): PV9138, dorsal rib. (D): SMM P84.12.26, anterior zygapophysis. (E): PV8830, pre-sacral diapophysis. Scale bars: A–D = 1 cm; E = 5 cm.

Site 6, Muldrow's Mill—ChM PV8850, PV8851, osteoderm fragments (not figured here).

Site 8, Burches Ferry—ChM PV7559, proximal part of femur.

Site 11, Kingstree—SMM P84.12.26, zygapophysis; SMM P84.12.28, distal portion of a left tibia.

DISCUSSION

A large number of indeterminate crocodilian bones are present in the overall assemblage, including a significant portion of limb bones. Femoral fragments include a distal portion, SCSM 2005.11.2 (see Figure 4.21C), which preserves part of the ridge for the adductor muscle, indicating a prominent fourth trochanter. A second crocodilian femur fragment, ChM PV7559 (see Figure 4.21D), differs from the other by the fourth trochanter located lower on the shaft; it shows relatively large curvature of the proximal end and it lacks the ventral muscle scar near the fourth trochanter that would be characteristic of *Thoracosaurus* (Mook 1931; Carpenter 1983).

SMM P84.12.28 (Figure 4.22A–C) is the distal half of a left tibia from a mid-sized crocodilian. The shaft is relatively straight; the midshaft is subcircular in cross-section, becoming triangular distally. The articular surface for the astragalus is broadly basinal, with a gently reflexed perimeter, and the contact surface for the fibula is a smooth depression.

Two additional anterior limb elements are those of hatchling-sized individuals. A right coracoid (PV8894, Figure 4.22D) and a right humerus (PV8895, Figure 4.22E) lack taxonomic characters due to their size and ablation of both ends of PV8895, the distal end of PV8894, as well as its proximal end below the coracoid foramen. Their presence at the site suggests the possibility that nesting sites of crocodilians were located within the general area. Another fragment of a young individual (PV9151, Figure 4.22F) is the left part of a frontal having partial contact sutures for the left prefrontal, postorbital, and parietal. This fragment indicates the presence of a "yearling-sized" individual or possibly a small taxon.

Skull material includes ChM PV8593 (Figure 4.23A), a cranial fragment consisting of a basisphenoid and basioccipital without its occipital condyle. This specimen is from an adult skull and is distinguished from *Borealosuchus* by the short, wide posterior surface of the basioccipital, which more nearly resembles that of *Alligator* in its widely flared basioccipital. The basisphenoid rostrum, preserved in part, is unusual in that its ventral edge is broad and flattened horizontally. A remnant of a hypophysial fossa is also present above the base of the rostrum anterior to the carotid canals. A tall, laterally compressed basisphenoid rostrum, normally exposed anterior to the braincase, is not present (Iordansky 1973).

A parietal (ChM PV8848, Figure 4.23B) represents an individual distinguished by a long frontal suture, slightly elevated margins of the supratemporal fenestrae, and a surface with large, deep, irregular pits. ChM PV6826

(Figure 4.23C) is a basisphenoid/basioccipital complex lacking diagnostic features. ChM PV8866 (Figure 4.23D) is a jugal with greater dorsoventral depth than in *Thoracosaurus,* but is otherwise nondiagnostic. A small, well-preserved articular (PV8858) shows numerous, fine lineations on external and internal surfaces, which may be selachian bite marks. This specimen is discussed subsequently in the discussion on bite marks on specimens in the assemblage. Fragmentary crocodilian skull and jaw bones lacking sufficient specific characters include two angulars (ChM PV8839 and PV8878), an angular/surangular (PV9140), two partial left quadrates (ChM PV8886 and PV9096), and a section of right dentary (SCSM 2006.10.1; not figured here).

Sixteen cervical and dorsal vertebral fragments, not figured here, cannot be specified taxonomically, as is typical of eusuchian vertebrae; however, the robustness of most of the specimens suggests they come from a larger taxon, such as *Deinosuchus* or *Thoracosaurus.* These include ChM PV7365, PV8676, PV7561, PV8312, PV8836, PV8837, PV8840, PV8841, PV8863, PV8864, PV8865, PV8882, PV8883, PV8884, PV8897, and PV9097. Two cervical ribs, PV8854 and PV8872 (Figure 4.24A, B, respectively); a dorsal rib, PV9138 (Figure 4.24C); a large single anterior zygapophysis, SMM P84.12.26 (Figure 4.24D); and a large presacral diapophysis, PV883 (Figure 4.24E), are among the taxonomically nonspecific axial skeletal specimens.

DINOSAURIA Owen, 1842
SAURISCHIA Seeley, 1887
THEROPODA Marsh, 1881
COELUROSAURIA Huene, 1914
ORNITHOMIMOSAURIA Barsbold, 1976
Ornithomimosauria indet.
Figure 4.25

MATERIAL

Site 1, Stokes Quarry—PV7558, manual ungual phalanx; PV8823, femoral fragment; PV8824, femoral fragment; PV8825, pedal ungual phalanx; PV9098, metatarsal fragment; PV9099, metacarpal.

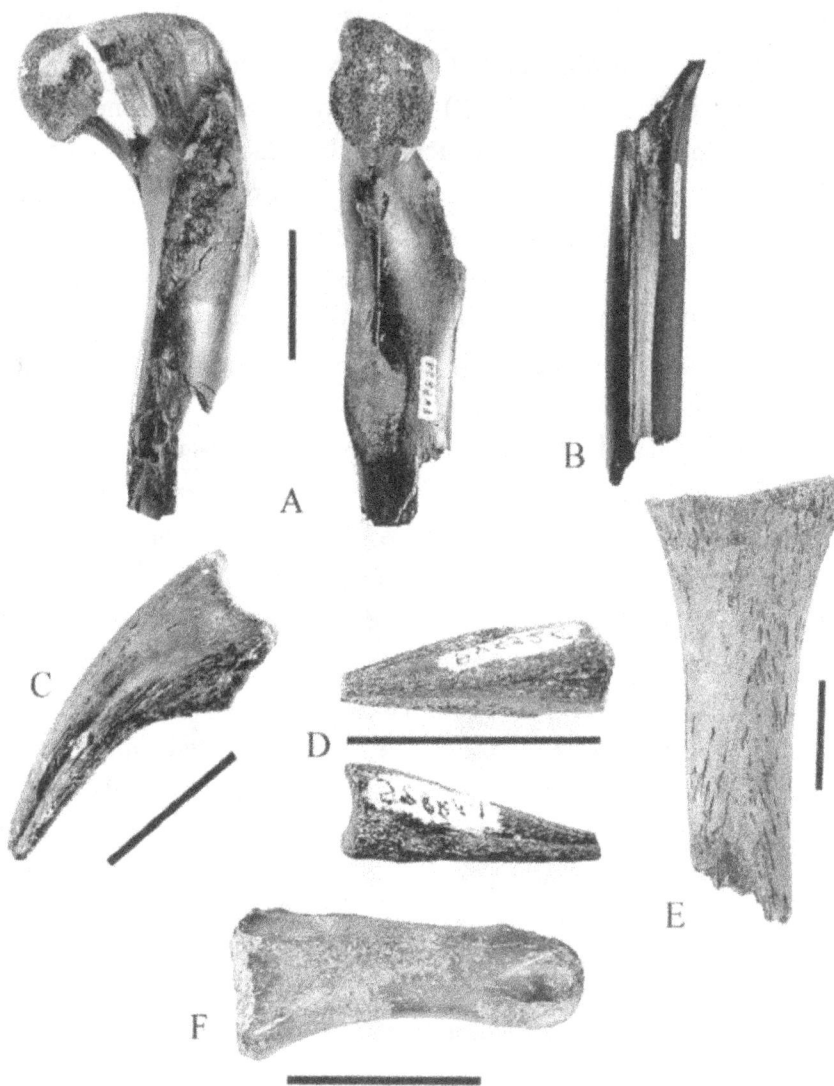

Figure 4.25 Postcranial bones of indeterminate ornithomimosaur dinosaurs. (A): ChM PV8824, proximal third of left femur. (B): PV8823, limb shaft fragment. (C): PV7558, manual ungual phalanx. (D): PV8825, pedal ungual phalanx. (E): PV9098, proximal half of metatarsal. (F): PV9099, metacarpal. Scale bars = 3 cm.

Site 11, Kingstree—ChM PV7530, PV7531, PV7300, PV7603, fragments of phalanges.

DISCUSSION

Ornithomimosaurs are a derived clade of long-armed, birdlike, coelurosaurian theropods, distinct from the maniraptorans by, among other features, a usual lack of teeth and the hypertrophied second pedal unguals of "raptors." They are farther characterized (Russell 1972; Makovicky, Yoshitsugu, and Currie 2004) by their gracile, elongate skulls, long hands with relatively straight claws, long hind limbs, and a generally slender, ostrich-like body plan (hence the name "bird mimic"). The isolated limb-shaft bones are generally identifiable based on their light construction, with large medullary cavities and thin cortices (Schwimmer et al. 1993). The ungual phalanges tend to be quite straight on the pes, and, as noted, teeth are absent. A recent study by Zelenitsky et al. (2012), showed that at least some species were feathered, in common with the maniraptorans. Fragmentary ornithomimosaur specimens have been widely reported across the eastern Late Cretaceous outcrop (Langston 1960; Baird and Horner 1979; Baird 1986; Russell 1988; Schwimmer et al. 1993).

Among ornithomimosaur specimens in the study at hand, a proximal third of a left femur (PV8824, Figure 4.25A) is identified based on the morphology of the head, which is directed straight medially, with greater height than breadth, and not separated from the greater trochanter by periosteal bone. The cranial trochanter is separated from the greater trochanter by a notch (Makovicky, Yoshitsugu, and Currie 2004). Another specimen, a limb-shaft fragment (PV8823, Figure 4.25B), shows the smooth internal surfaces around the medullary cavity and thin cortex characteristic of ornithomimosaur bone. Both specimens show "fresh" fracturing that probably resulted from mining machinery rather than from erosion or transportation at the time of deposition; hence, these two specimens could well have come directly from the Campanian sediments at Stokes Quarry rather than from the unconformably overlying lag deposit at the base of the Duplin Formation at the quarry (see Figure 3.1A, p. 23).

PV7558 (Figure 4.25C) is a typical ornithomimosaur manual ungual, either the second or third, which is relatively long, and although slightly curved, does not show the distinctly talon-shaped hook characteristic of the raptorial claws in maniraptors and tyrannosaurs (discussion to follow). A characteristically straight pedal ungual (PV8825, Figure 4.25D), is also attributable to an ornithomimosaur. Other ornithomimosaur podials include ChM PV9098 (Figure 4.25E), the proximal half of a metatarsal that resembles metatarsal II of some ornithomimosaurs (Barsbold and Osmólska 1990; Makovicky, Yoshitsugu, and Currie 2004) in that it is hollow and expanded proximally with a smoothly concave articular surface, and a metacarpal (PV9099; Figure 4.25F) from Stokes Quarry, likely from an ornithomimosaur based on the gracile morphology.

MANIRAPTORIFORMES Holtz, 1996
Family DROMAEOSAURIDAE Matthew and Brown, 1922
Subfamily SAURORNITHOLESTINAE Longrich and Currie, 2009
Saurornitholestes Sues, 1978
Saurornitholestes langstoni Sues, 1978
Figure 4.26

MATERIAL

Site 1, Stokes Quarry—SCSM 2005.11.1, ChM PV8674, 8675, teeth; PV8679, pedal ungual phalanx.
Site 8, Burches Ferry— SCSM 99.55.1, pedal ungual phalanx.

DISCUSSION

Teeth from the velociraptorine *Saurornitholestes langstoni* have been previously identified in the eastern Late Cretaceous (Kiernan and Schwimmer 2004; DRS personal observation, 2012). The identification of dromaeosaur teeth is based on the distinctive morphology of the denticles (Currie, Rigby, and Sloan 1990; Sankey, Standhart, and Schiebout 2005) with extreme disparity in the size of the denticles on cranial (anterior) and

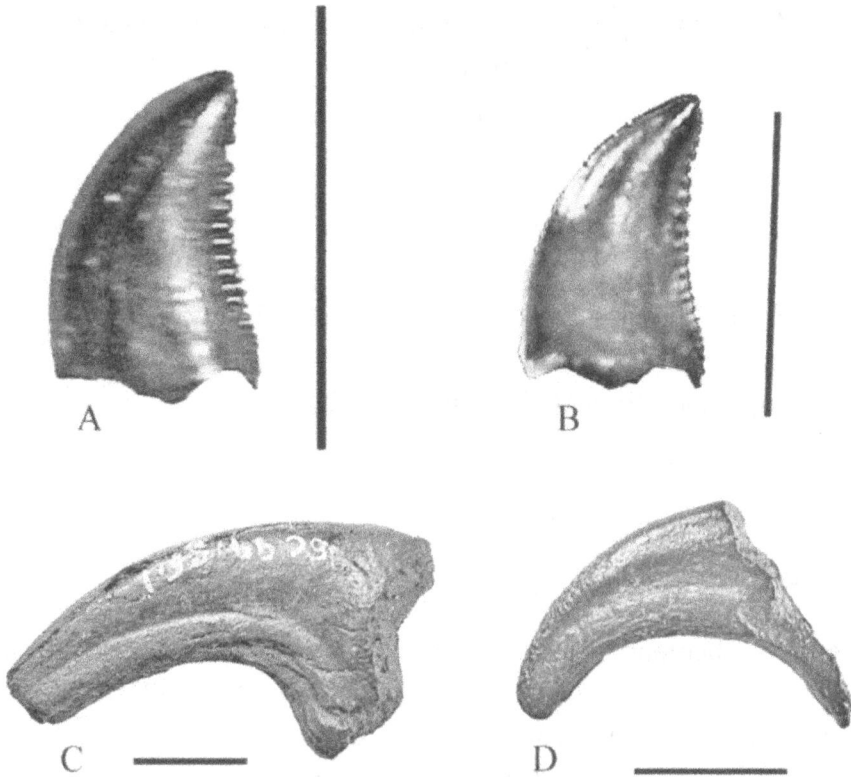

Figure 4.26 Teeth and ungual phalanges of the velociraptorine dinosaur *Saurornitholestes langstoni*. (A): SCSM 2005.11.1. (B): ChM PV8674, teeth. (C): SCSM 99.55.1. (D): PV8679, pedal ungual phalanges. Scale bars: A, B = 5 mm; C–D = 1 cm.

caudal (posterior) carinae. ChM PV8674 (Figure 4.26B) and PV8675 (not figured here) exemplify this morphology, with much larger, distinctly pointed denticles on the caudal carina, and with minute denticles toward the apex of the cranial carina, becoming nearly indistinct toward the base of the crown. These tooth crowns are approximately 1.0 centimeter or less in height, relatively thin laterally, and strongly hooked in lateral profile.

PV8679 and SCSM 99.55.1 (Figure 4.26C, D) are relatively small, talon-shaped ungual phalanges, with the deep lateral grooves typical of dromaeosaur pedal unguals. SCSM 99.55.1 is missing the sharp tip that might provide enough overall shape to show whether it is the hypertrophied second ungual phalanx characteristic of the "raptors."

TYRANNOSAUROIDEA Walker, 1964
Appalachiosaurus montgomeriensis Carr, Williamson and
Schwimmer, 2005
Figure 4.27

MATERIAL

Site 1, Stokes Quarry—ChM PV7326, PV8826, PV9117, teeth; PV7370, limb fragment.
Site 3, Turbeville—PV6819, large pedal phalanx.
Site 8, Burches Ferry—SCSM 98.64.2, tooth.

DISCUSSION

The largest theropod teeth in the assemblage range in crown height from 3.0 to 4.0 centimeters (as reconstructed in the case of PV 7326). The denticles are relatively small and are of similar size and shape on both cranial and caudal carinae. The denticulation is best observed in specimens PV7326 and PV8826 (Figure 4.27A, B). In contrast with the pointed denticles noted previously for the velociraptorine *Saurornitholestes* (see, e.g., Figure 4.27A), these larger teeth feature denticles with rounded, chisel-shaped tips, and relatively large spaces between the denticles. This denticle morphology might be termed "hatchet-head shaped" and is characteristic of tyrannosauroid teeth (Sankey Standhardt, and Schiebout 2005). Despite the common shape of denticles among the tyrannosauroids, the overall size of teeth, presence and size of denticles on the carinae, and the shape and positions of the carinae are highly variable among tyrannosaurs (Holtz 2004).

The notably small size of the denticles in these South Carolina specimens is consistent with the teeth of the monotypic tyrannosauroid genus *Appalachiosaurus*, which was originally described from a type specimen from central Alabama (Carr, Williamson, and Schwimmer 2005), and isolated teeth and bones from western Georgia (Schwimmer et al. 1993). The two teeth in the South Carolina assemblage lacking observable denticulation, PV9117 and SCSM 98.64.2 (see Figure 4.28C, D) are tentatively

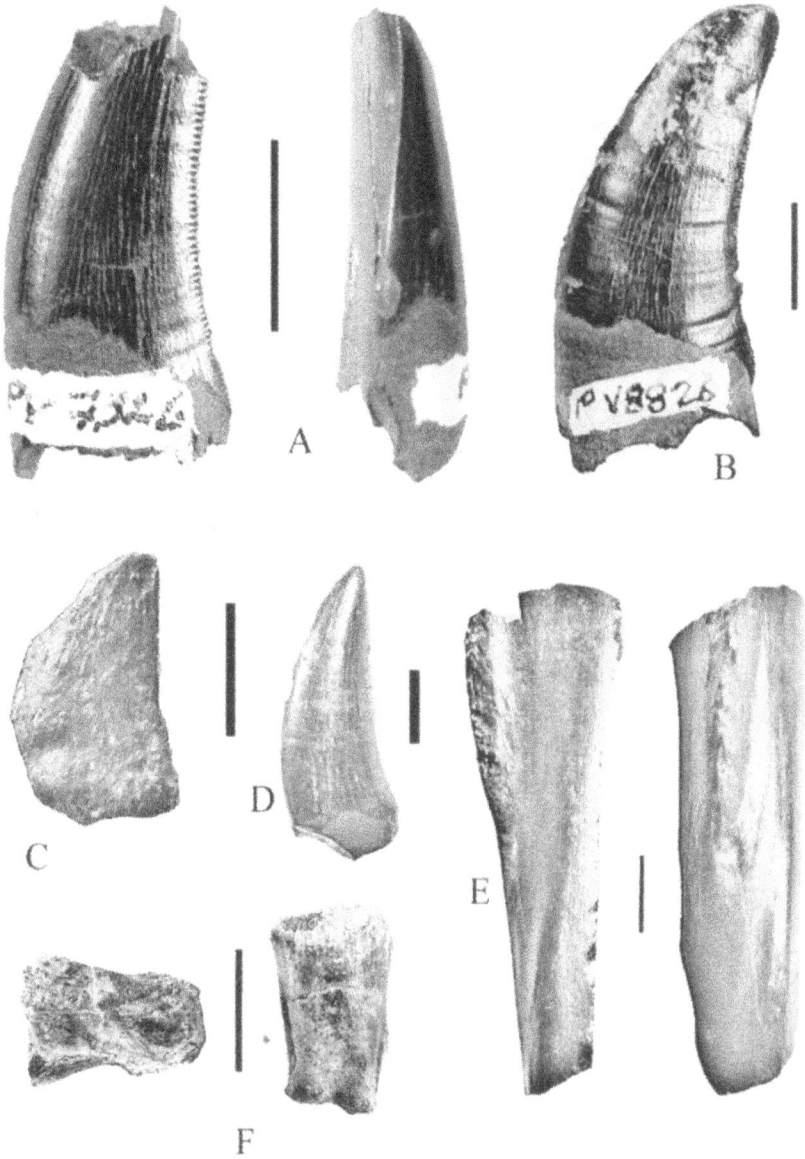

Figure 4.27 Teeth and postcranial bones of the tyrannosauroid dinosaur *Appalachiosaurus montgomeriensis*. (**A**): ChM PV7326. (**B**): PV8826. (**C**): PV9117. (**D**): SCSM 98.64.2; teeth. (**E**): PV7370, limb fragment. (**F**): PV6819, pedal phalanx. Scale bars: **A–E** = 1 cm; **F** = 3 cm.

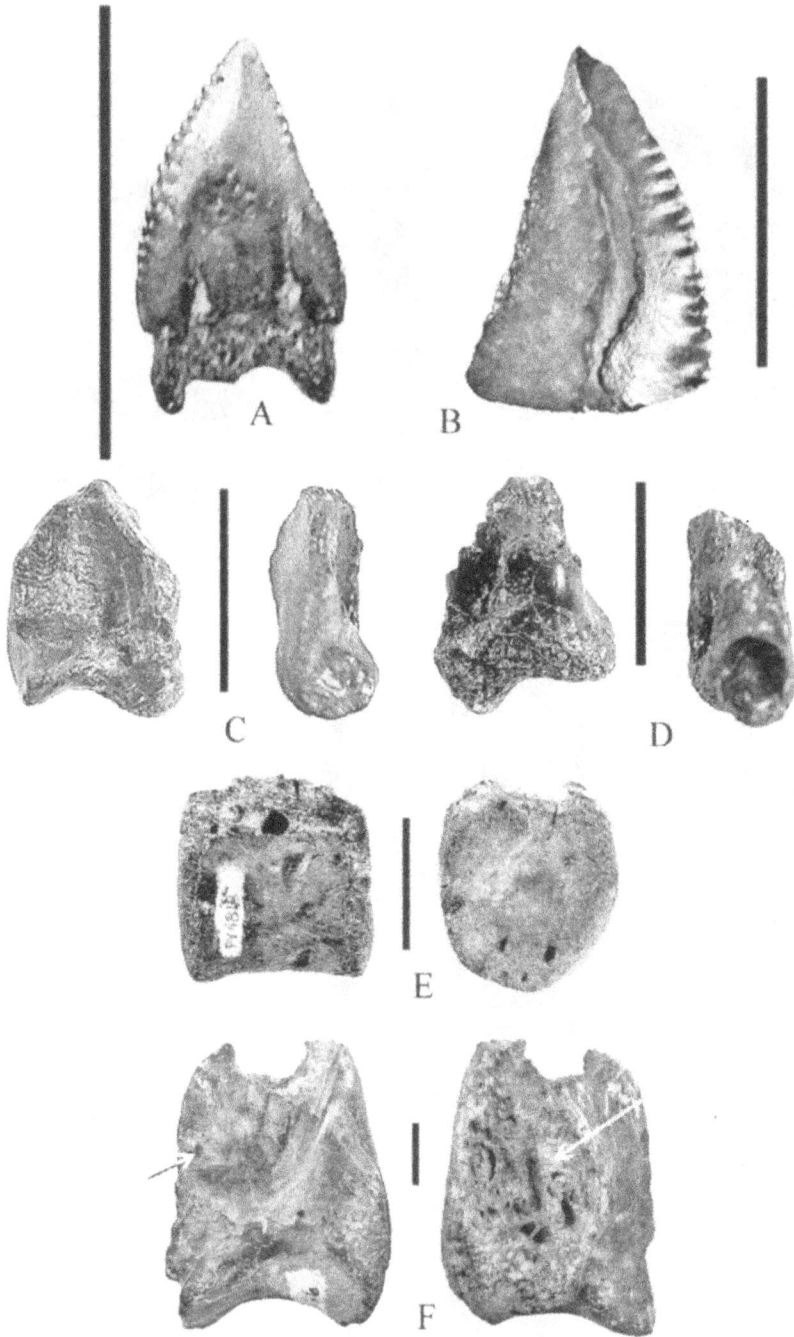

Figure 4.28 Teeth and bones of theropod dinosaurs. (A): ChM PV8689. (B): PV9111, teeth. (C): PV9149, distal end of humerus. (D): PV9150, distal end of tibia. (E): PV4818, dorsal vertebral centrum. (F): PV 7366, pathologic distal metatarsal fragment. Scale bars: A, B = 5 mm; C, D, = 1 cm; E, F = 3 cm.

assigned to *Appalachiosaurus* because of their larger size and overall shape. It is notable that all the tyrannosauroid teeth in this assemblage are within the size range of *Appalachiosaurus* teeth known from Georgia and Alabama (DRS personal observation) because the holotype and referred specimens are considered to be juveniles based on a number of osteological character-istics (Carr, Williamson, and Schwimmer 2005). By extension, this sug-gests that the teeth here referred to *Appalachiosaurus* are also from juvenile individuals.

A single limb fragment, PV7370 (Figure 4.27E), is tentatively assigned to *Appalachiosaurus* because of the relatively robust cortical tissue and smooth medullary cavity lining, suggesting a heavier bodied theropod. However, there is not sufficient material to further diagnose the specimen. Similarly, a larger pedal phalanx, PV6819 (Figure 4.27F) is of comparable size to the pedal phalanges of *Appalachiosaurus*, but it is insufficiently preserved to see whether the characteristic "lipping" of the cranial–dorsal aspect is present (Carr, Williamson, and Schwimmer, 2005).

Theropoda indet
Figure 4.28

MATERIAL

Site 1, Stokes Quarry—ChM PV8689, PV9111, teeth; PV 7366, pathologic manual phalanx, PV 8833, limb shaft fragment; ChM PV9149, PV9150, distal limb fragments, humeral or tibial; ChM PV7366, metatarsal fragment.
Site 14, near Little River— ChM PV4818, vertebral centrum.

DISCUSSION

ChM PV8689 (Figure 4.28A is a minute theropod tooth crown (approxi-mately 3.0 mm high) with larger denticles on the cranial carina, a relatively convex lingual side, and a subtriangular lateral profile. This morphology is sufficiently distinct from *Saurornitholestes* to exclude that genus as the identity, but it does not allow a firm placement in other taxa. Several small, contemporary maniraptoran genera have teeth that resemble PV 8689, including *Troodon* and *Dromaeosaurus* (Currie, Rigby, and Sloan 1990), but neither genus has been previously identified in the eastern Cretaceous outcrop, and this tooth is not sufficient material to designate a significant range extension. A second indeterminate theropod tooth crown, PV9111 (Figure 4.28B), shows similar denticles on what appears to be the cranial carina, however, it is too ablated to be certain of the overall shape and the nature of the presumed posterior cranial surface. Both teeth, although in part poorly preserved, are not obviously hooked posteriorly.

ChM PV9149 and PV9150 (Figure 4.28C, D) are distal ends of a bird-like humerus and tibia, respectively. They compare favorably with the same elements of many smaller theropods and cannot be assigned more precisely than Maniraptora? indet. A platycoelous dorsal vertebral centrum (PV4818, Figure 4.28E), is tentatively regarded as a theropod centrum because of its proportions, with its width less than its height, and its height less than its length. A small, shallow pit located at the base of the pedicel, about midlength of the centrum, is similar to that found on some ornithomimid vertebrae (Barsbold and Osmólska 1990). Several small indentations on the sides and ventral surface may be tooth marks.

PV7366 (Figure 4.28F) is the distal end of a metatarsal with a hypertrophic condition of the shaft indicating overgrowth of bony tissue just above the articular facet on the flexor side and erosion on the extensor side.

Several aspects of the occurrences of Late Cretaceous theropod remains in the deposits on the Coastal Plain of South Carolina are noteworthy. First, the simple number of theropod specimens, despite the uncertain identifications resulting from the typically poor condition, suggests that predatory dinosaurs of all sizes were abundant, and probably relatively diverse, given at least four different taxa identified (including the indeterminate tooth specimen, PV8689, which is nevertheless clearly distinguishable from the other theropods). Second, it is notable that no very large theropod individuals are represented, especially observing that the largest species in the assemblage, *Appalachiosaurus montgomeriensis*, is represented only by immature specimens. This observation corresponds to similar ones noted by Schwimmer (2002), who hypothesized that the large predator niche along most of the Late Cretaceous Coastal Plain in eastern North America was occupied by the giant crocodilian *Deinosuchus rugosus*, rather than by tyrannosaur dinosaurs, as is the case in the contemporary western portion of the continent. And indeed, although *Deinosuchus* material is sparse in South Carolina, teeth in the assemblage are from typical-size individuals.

ORNITHISCHIA Seeley, 1887
ORNITHOPODA Marsh, 1881
Family HADROSAURIDAE Cope, 1869
Hadrosauridae genus indet.
Figures 4.29, 4.30A–D

MATERIAL

Site 1, Stokes Quarry—ChM PV8827, vertebral centrum; PV8828, tooth fragment.
Site 11, Kingstree—SCSM 87.158.1, 87.158.2, 87.158.51, PV7676, tooth fragments; SMM P84.12.27, exoccipital fragment; ChM PV7584, maxillary fragment.

DISCUSSION

Hadrosaur remains are widely reported from Upper Cretaceous outcrops throughout the eastern United States (Leidy 1858; Lull and Wright 1942; Langston 1960; Baird and Horner 1979; Schwimmer 1997a; Schwimmer et al. 1993), and are often found in marine deposits with other, possibly coastal, nonmarine taxa, such as lizards, turtles, and crocodilians. Nevertheless, few eastern hadrosaur remains can be classified below the family level because the diagnostic skull and pelvic bones are rarely preserved, whereas teeth, mandibles, distal vertebrae (i.e., caudals) and distal limbs are common. A detailed explanation of this is presented in Schwimmer (1997a), and relates to near-shore marine taphonomic processes, and scavenging by the shark species *Squalicorax* (see discussion to follow

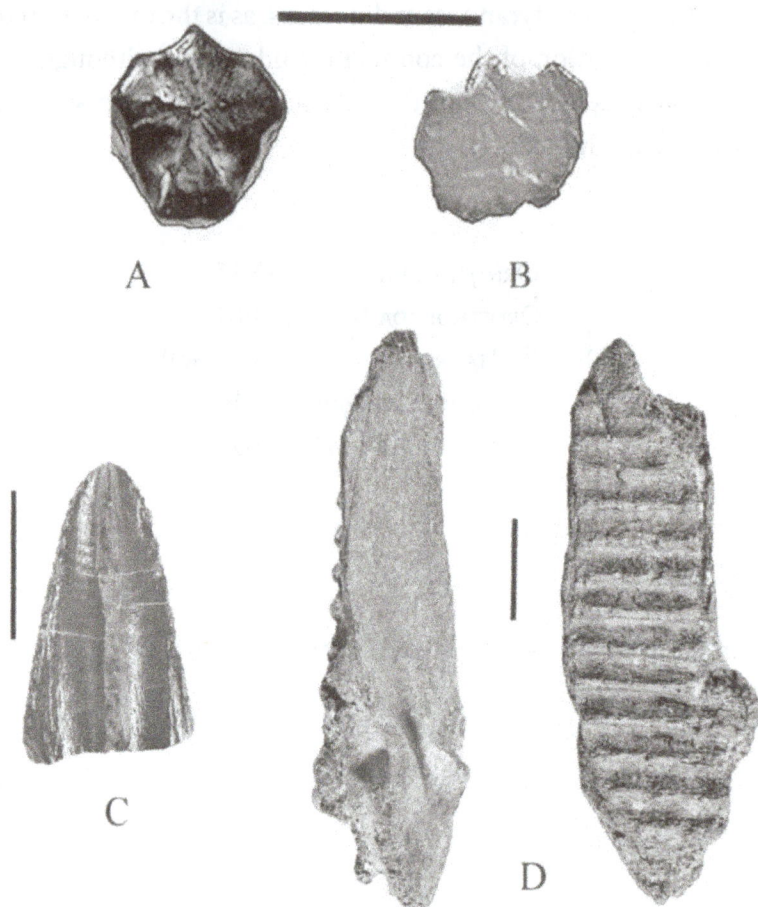

Figure 4.29. Hadrosaur teeth and maxilla. (A–C): Teeth. (A): SCSM 87.158.1. (B): 87.158.2, sectional views. (C): ChM PV8828, lingual view. (D): PV7584, maxillary fragment. Scale bars: A–C = 5 mm; D = 1 cm.

about bite-marked bones). The sole example in the existing literature of an eastern hadrosaur with ample cranial material is the juvenile holotype of the hadrosaurine *Lophorhothon atopus* (Langston 1960). This specimen from the western Alabama Mooreville Chalk is frequently considered the reference taxon for many southeastern hadrosaur finds (e.g., Schwimmer et al. 1993); however, because it is a juvenile, comparisons with adult specimens are tenuous.

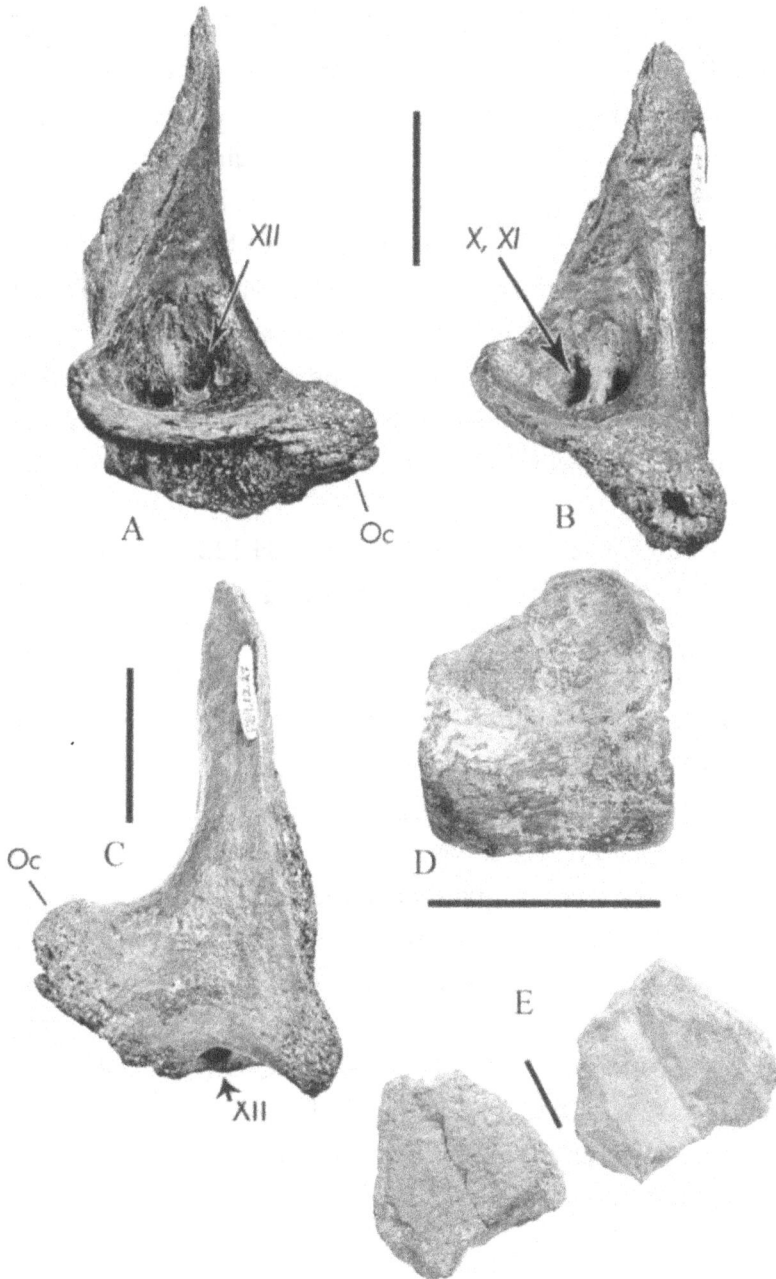

Figure 4.30 Hadrosaur and indeterminate dinosaur cranial fragments and vertebra. (A–C): SMM P84.12.27, hadrosaur partial left exoccipital. (A): Lateral view showing foramen for cranial nerve XII. (B): lateral oblique view showing foramen for cranial nerves X and XI. (C): foramen for cranial nerve XII; Oc = occipital condyle. (D): ChM PV 8827, hadrosaur ablated dorsal or mid-caudal vertebral centrum, lateral view. (E): PV9143, probable parietal of small, indeterminate dinosaur. Scale bars: **A–C, E** = 3 cm; **D** = 5 cm.

Among the hadrosaurid remains from the Kingstree site included here are an incomplete left exoccipital (SMM P84.12.27; Figure 4.30A–C) collected in 1984, and two teeth (SCSM 87.158.1 [Figure 4.29A] and 87.158.2 [Figure 4.29B]), collected in 1986, that were reported by Weishampel and Young (1996) as the first records of dinosaurs found in South Carolina. These were not particularly diagnostic, but among the material discovered subsequently are additional teeth, ChM PV8828 and PV7676, and two skull fragments: a small maxilla (PV7584; Figure 4.29D) collected in 2007, apparently from a juvenile hadrosaur, and a partial left exoccipital (SMM P84.12.27; Figure 4.30A–C). The well-preserved tooth crown (PV8828; Figure 4.29C) shows fine denticulation in the area best preserved near the apex, which is characteristic of the hadrosaurine subfamily (Langston 1960).

The incomplete left exoccipital (SMM P84.12.27; Figure 4.30A–C) preserves part of the paroccipital process and a portion of the occipital condyle, and houses foramina for cranial nerves X, XI, and XII. Sutures that unite the exoccipital, opisthotic, and basioccipital bones evidently fused in early development (Lull and Wright 1942; Ostrom 1961; Horner 1992), and associations with the present exoccipital are difficult to distinguish. According to Romer (1956) and Ostrom (1961) the hypoglossal nerve, the twelfth cranial nerve, may emerge from within the skull via one, two, or three foramina, partly confluent or independent of the vagus nerve, cranial nerve X. In the present specimen, nerve XII emerged from a single foramen. Nerve X and XI emerged through partly confluent foramina.

ChM PV8827 (Figure 4.30D), from Stokes Quarry, is an ablated, small, vertebral centrum with the typical shape of a hadrosaur dorsal or midcaudal vertebra; however, in the absence of the neural region and in the state of preservation, the position is unclear.

Dinosauria? indet.
Figure 4.30E

MATERIAL

Site 1, Stokes Quarry—ChM PV9143, skull fragment? PV8956, possible distal fibula (not figured here).

DISCUSSION

A cranial fragment (PV9143; see Figure 4.30E) resembles the parietal region of a small dinosaur, possibly a theropod. There is an apparent midline suture, indicating that it comes from the center of the skull. The absence of pitting on the external surface excludes a crocodilian identity, but there is not enough information to further classify the specimen.

PV8956 is a distal limb-shaft fragment, with many ablated surfaces. The internal structure is characteristic of many tetrapods, excluding theropods, and the apparent distal end shows the shape characteristic of a small hadrosaur fibula. However, as with the cranial fragment described previously, there is insufficient detail to reliably classify the specimen.

5

TRACE FOSSILS

Coprolites
Figures 5.1, 5.2

MATERIAL

Site 1, Stokes Quarry—ChM PV8697, PV8698, PV8999, PV9127: crocodilian coprolites without bone inclusions; PV9000, PV9154: selachian coprolites with small bone, tooth, and/or scale inclusions; PV8998: selachian coprolite with well-preserved vertebrae inclusions.
Site 7, Diamondhead Loop Road—ChM PV8997: crocodilian coprolite.
Site 11, Kingstree—ChM PV7562: crocodilian coprolite without bone inclusions.

DISCUSSION

Identification of a coprolite's source is always tentative, except for the most stereotypic forms, such as selachian coprolites with unambiguous evidence of their spiral intestinal structure (Williams 1972; Stewart 1978; Stringer and King 2012). The majority of coprolite specimens in the present study appear to be from crocodilians and selachians, which conforms to the dominance of crocodilian bones in the tetrapod assemblage and the general abundance of selachian teeth (notably of the genus *Squalicorax* [Cicimurri 2007]) in the deposits.

Recent work on crocodilian coprolites and feces (Harrell and Schwimmer 2010; Souto 2010) shows that a variety of shapes are possible, but among the most characteristic are cylindrical to tapering shapes (Hunt and Lucas 2010), often with slight axial twisting, lineations and convolutions on the external surface, and terminations with one or more pointed tips. Among the internal characteristics of crocodilian coprolites, based in part on studies of modern feces (Milàn 2010), are gas-bubble voids and the absence of recognizable bone fragments, which is consistent with the results caused by the strong gastric juices of crocodilians.

Figure 5.1 Coprolites assumed to be from crocodilians, lacking evident bone inclusions. (A): ChM PV8999. (B): PV8698. (C): PV9127. (D): PV7562. (E): PV8997. Scale bars: A–C, E = 30 mm; D = 50 mm.

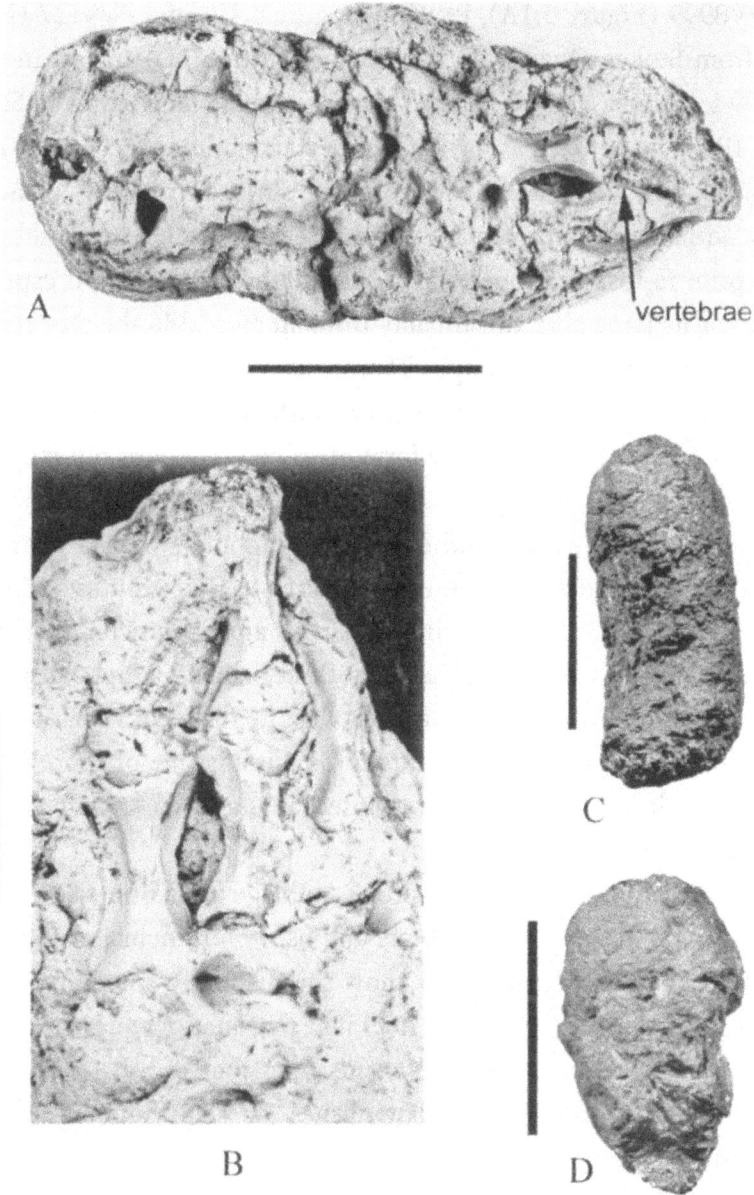

Figure 5.2 Coprolites of sharks. (A): ChM PV8998, specimen with inclusions of cervical vertebrae of hatchling trionychid turtle. (B): enlarged view of posterior end ChM PV8998 showing four trionychid cervical vertebrae. (C): PV9000. (D): PV9154. Scale bars: A = 10 mm; B = 3 mm; C, D = 20 mm. A and B were whitened with ammonium chloride sublimate.

Coprolites that appear to be of crocodilian origin are abundant at Stokes Quarry and present at Kingstree. Most characteristic of these are ChM PV8999 (Figure 5.1A), PV8696 (Figure 5.1B) and PV9127 (Figure 5.1C) from Stokes Quarry, and PV7562 (Figure 5.1D) from Kingstree. PV 8999 features a notably twisted shape that matches figures in Sawyer (1981, 1998) recognized as having a typical crocodilian fecal form. Additionally, Sawyer (1998) noted the prevalent "surface wrinkling" of ostensible crocodilian coprolites from the Kingstree site, which he attributed to drying prior to fossilization. PV7562, although fragmentary, is especially notable for its large size, presumably from an animal in the size range of a smaller *Deinosuchus* or a larger *Thoracosaurus*.

Two coprolites from Stokes Quarry (ChM PV8698, PV9127; Figure 5.1B and C) differ from the usual form of crocodilian feces but fall within the possible shapes of many larger carnivores, including crocodilians (Hunt and Lucas 2010). ChM PV8698 is globular, with a prominent "nipple-like" termination, common on terrestrial coprolites (Waldman and Hopkins 1970), and bearing apparent anal sphincter marks and surface striae. ChM PV9127 consists of three stacked, elliptical sections, flattened on one surface as if a soft mass were deposited on a hard surface. This oddly shaped structure appears to be missing one or more sections on one end, has no apparent bone internally, and contains a few very small gas cavities.

Many coprolites in the composite assemblage are assumed to be selachian in origin because of their bone and fish-tooth inclusions, but none are sufficiently well preserved to show the distinctly spiral structure of ideal shark specimens (Williams 1972). The presence of bone in a coprolite indicates a less acidic digestive systems than that of, say, crocodilians, and is typical of both sharks and bony fishes. These presumably selachian specimens include PV9000 and PV9154 (Figure 5.2C, D), both from Stokes Quarry, each consisting of smaller (~3.0 cm), tapering masses with extensive surface wrinkling and bone fragments on the external surfaces. It is likely that the remaining surface structure is a vestige of the original spirals lost during fossilization of the softer masses. Most notably in PV9154 are many osteichthyan tooth fragments and an apparent teleost vertebra.

PV8998 (Figure 5.2A), also a small (3.4 cm) shark coprolite from Stokes Quarry, contains inclusions of six, well-preserved, elongate, opisthocoelous vertebrae with long zygapophyses, and visible parts of two more vertebrae. These vertebrae are clearly in association and identifiable as the cervical series of a very small trionychid turtle. The vertebrae are oriented in two approximately parallel series (Figure 5.2B), with the ventral surfaces on the external portion of the coprolite. This coprolite is tentatively attributable to a shark, with slight evidence of spiral surface texture, and it implies feeding by a small shark on an even smaller freshwater turtle hatchling that may have drifted downriver to the shore face.

Bite Marks
Figure 5.3

MATERIAL

Site 1, Stokes Quarry—PV8856, crocodilian radius fragment; PV8857, indeterminate manual carpal (?) fragment; PV8858, crocodilian articular with preserved retroarticular process; PV8995, turtle scapular fragment. *Site 14, near Little River*—ChM PV4818, ornithomimosaur? vertebral centrum.

DISCUSSION

A number of specimens from Stokes Quarry exhibit bite marks attributed to the lamnoid shark genus *Squalicorax*, which left abundant teeth in the assemblage, making up about 20 percent of the total recovered shark material. *Squalicorax* species were apparently the preeminent marine scavenger of vertebrate carcasses during the Santonian and Campanian ages of the Late Cretaceous (Schwimmer, Stewart, and Williams 1997), leaving characteristic parallel rows of denticle marks, often with a single puncture from the apex of the tooth. They are represented in the present fauna by at least two species, *S. pristodontus* and *S. kaupi* (Cicimurri 2007). The

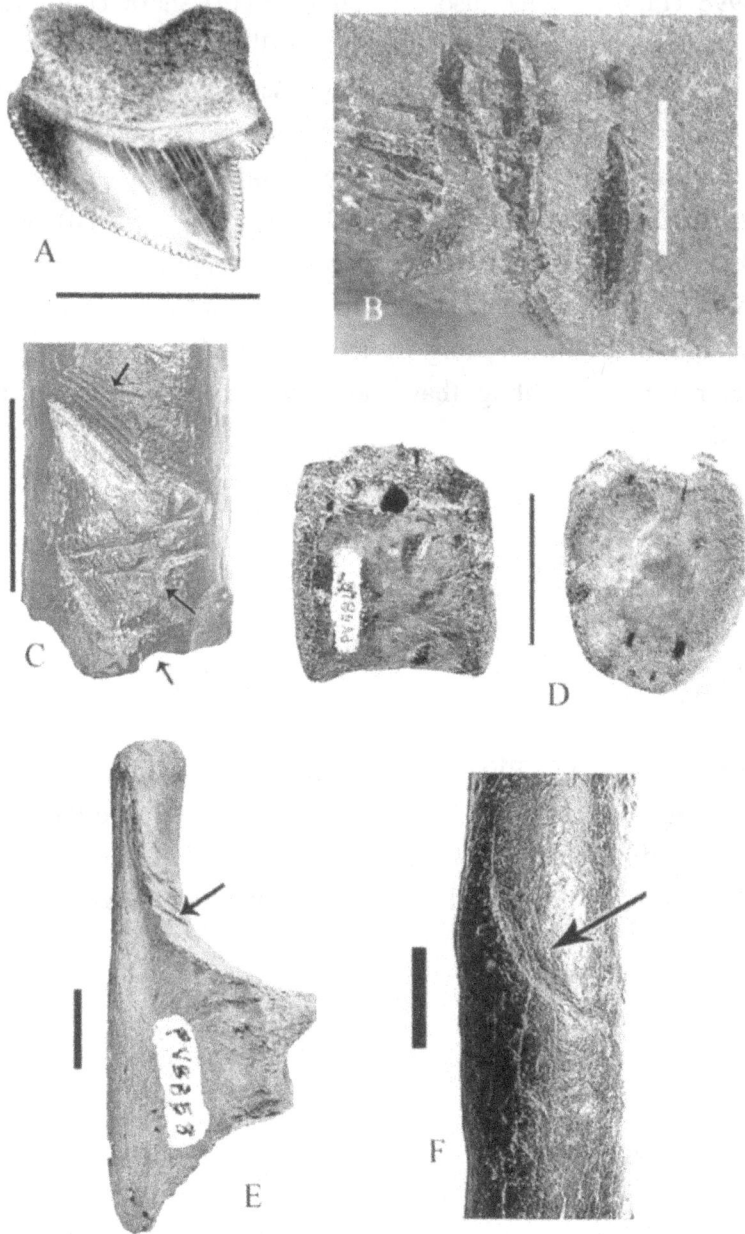

Figure 5.3 Shark and other bites on bones. (A): SMM P2006.2.1 *Squalicorax kaupi* tooth. (B): ChM PV8995, fragment of turtle scapula with *Squalicorax* tooth marks. (C): PV8857, theropod shaft fragment with *Squalicorax* tooth marks: both B and C show striations from shark tooth denticles. (D): PV4818, vertebral centrum with round crocodilian tooth marks. (E): PV8858, crocodilian retroarticular with indeterminate bite marks. (F): PV8856, crocodilian radius fragment with indeterminate bite mark. Scale bars: A, E, F = 1 cm; B, C = 5 cm; D = 3 cm.

Squalicorax teeth here range in height from 6 to 22 millimeters: Figure 5.3A is a tooth of *Squalicorax kaupi* from the assemblage SMM P2006.2 (lot of forty-five specimens).

A fragment from the base of a turtle scapula (ChM PV8995; Figure 5.3B) has eight or more *Squalicorax* bite marks made on fresh bone, as shown by the deep penetration and the raised periosteal surfaces adjacent to the tooth impressions. Several of the impressions show punctures caused by the sharply pointed apices of individual *Squalicorax* teeth. Considering that the scapula is situated well inside the shell of a turtle, it is unlikely that those bites were inflicted while the animal was living; thus, these are evidently scavenging traces.

A small, indeterminate bone fragment (ChM PV8857; Figure 5.3C), possibly a theropod or crocodilian metapodial composed of dense cortical bone with a tiny medullary cavity, also bears the characteristic denticulate, incised marks of *Squalicorax* at both ends. These examples of scavenging or feeding by *Squalicorax* attest to a marine environment in which floating carcasses afforded opportunities for shallow-marine or fluvial-transported carcasses to reach marine predators and scavengers (Langston 1960; Schwimmer 1997a).

Evidence of crocodilian predation is most evident in a vertebral centrum (ChM PV4818; Figure 5.3D) possibly from an ornithomimid dinosaur (as previously discussed). This well-preserved centrum bears many blunt, subcircular, shallow pits. Most significant, these pits are located on three surfaces (two lateral, plus one articulating end) suggesting the vertebra was bitten on one end by a relatively small animal with powerful jaws and blunt teeth. Crocodilians are notable for both attributes, and for leaving blunt tooth marks on bones (Njau and Blumenschine 2006; Schwimmer 2010).

Specimens showing bite marks of uncertain origin include a small, complete crocodylian right articular (ChM PV8858; Figure 5.3E). This specimen has long, finely incised tooth marks on the dorsomedial aspect of the retroarticular process that abruptly end at the sutural border of the missing surangular. There are also finely incised lines on the internal surfaces of the articular, possibly bite marks, suggesting that the bone was separated from the remaining jaw and well chewed during this feeding episode. The articular presents little indication of other erosion and may have belonged to a floating carcass when fed on by a scavenger. The bite marks lack evidence of denticulation, therefore, they could have originated from a number of smaller marine sharks or bony fishes. Likewise, PV8856 (Figure 5.3F), a section of the shaft of a small crocodilian radius, exhibits one long, curved scar, of undetermined origin.

6

TAPHONOMY and PALEOECOLOGY

Late Cretaceous deposits at fourteen locations in the South Carolina Coastal Plain were investigated during the course of this study (see Figure 1.1, p. 4). These are mostly lag accumulations containing highly concentrated remains of vertebrate fossils, which comprise the most common type of occurrences of Cretaceous vertebrates on the East Coast (Miller 1968; Schwimmer 1986, 2002). Of approximately 450 fragmentary tetrapod specimens collected for this study, 275 are included in the present analysis to provide a comprehensive representation of the reptilian and dinosaurian faunal composition. The relative abundances of fossils by taxa at the various sites can be seen in Table 1.1 (p. 6). Specimens from these sites share taphonomic features indicating that various degrees of transportation were involved before final deposition in the near-shore marine environments. Baird and Horner (1979) related the heterogeneous nature of the coeval Black Creek vertebrate fauna of North Carolina to the varied circumstances of deposition, which included fluvial, fluviomarine, lagoonal, and littoral origins.

According to Schwimmer (2002), highly concentrated eastern Late Cretaceous coastal deposits are the result of a cyclic process of sea-level change by which carcasses and bones from streams and estuaries were transported into deeper water during a regressive phase of the sea-level cycle and subsequently retransported back up onto the marine shelf during a transgressive phase. With each sequence of transport, further abrasion and concentration of specimens occurs (Rogers and Kidwell 2007, 25–26). In the Pee Dee River region of South Carolina, which yielded the greater portion of our material, the Cretaceous deposits that comprise the Black Creek Group were deposited in the outer regions of fluvial sedimentary wedges prograding into the marine environment (Prowell et al. 2003). The cyclic deposition model (*fide* Haq, Hardenbol, and Vail 1987) further accounts for the admixing of open-marine to nonmarine taxa into the near-shore marine setting. That concept supports the observation that the highly concentrated assemblage from Stokes Quarry, for example, ranges from fragmented bones and teeth that show severe wear, with all surface detail gone and bones that cannot be reliably identified, to specimens that show little taphonomic modification from wear. Some of these elements, however, have sharp angular fractures that resulted not from transport but from subsequent excavation by machinery:

for example, the ornithomimosaur femur ChM PV8824 (see Figure 4.25A, p. 94).

The setting along the extensive, low, Coastal Plain of South Carolina of the Late Cretaceous Period, was probably one of salt marshes, tidally influenced estuaries, and wide intertidal zones, frequented by terrestrial, estuarine, and shallow-water marine tetrapods. A comparable study of fishes in the upper Campanian Donoho Creek Formation by Cicimurri (2007), confirms the general dominance of nearshore, neotropical marine sharks and osteichthyes, admixed with some terrestrial fossils. Cicimurri follows Prowell et al. (2003) in recognizing the general style of sedimentation of the Black Creek Group as deposition in a prograding delta system, with the Donoho Creek Formation comprising the delta front to pro-delta facies.

The abundance of freshwater turtles, fluviatile and estuarine crocodilians, and dinosaurs in the collective South Carolina Cretaceous assemblage shows that there was a significant terrestrial vertebrate presence in the nearshore marine environment. However, in assessing the overall situation represented by the taxa studied here, it is important to note the large component of marine tetrapods (sea turtles, mosasaurs and plesiosaurs) as well as the abundant remains of marine bony fishes and selachians not discussed here. The abundance of marine taxa in the assemblage clearly shows that these are fundamentally marine deposits, as transportation of terrestrial carcasses into marine environments is a commonplace occurrence. In addition to the fishes reported by Cicimurri (2007), among the amateur collections from Stokes Quarry, the private collections of R. Ogilvie and F. Morning contain abundant marine bony fish, including *Anomoeodus phaseolus*, *Enchodus petrosus*, and *Hadrodus priscus*, as well as specimens of many marine reptile taxa and abundant teeth of sharks, especially *Squalicorax kaupi*.

And yet, it is not unusual for modern marine vertebrates to travel for some distance into upper estuarine waters. Bottlenosed Dolphins (*Tursiops truncatus*) are a common sight in the estuaries and tidal creeks of coastal South Carolina, coauthor Albert E. Sanders having seen them in Goose Creek near its juncture with the Cooper River above Charleston, at least 16 kilometers upstream from the ocean. Though one must be very careful in drawing conclusions about the behavioral patterns of extinct, ancient taxa from those of modern animals, there is no reason to believe that Cretaceous marine vertebrates did not frequent the relatively safe and quiet waters of upper estuaries. Also relating the terrestrial fauna to marine fauna, the occurrence of several types of dinosaurs and four taxa of crocodilians in the overall assemblage suggests the presence of well-sorted, well-drained fluvial sands available for nesting sites in proximity to the marine shoreline, which may have been remote from the adult home range (Coombs 1990).

The admixture of remains of both marine and nonmarine vertebrates in sedimentary deposits in the Eastern Upper Cretaceous is common, and along with the fragmentation and concentration of bones and teeth, is evidence of marine reworking of older materials into overlying deposits. For example, at the Quinby locality (Site no. 2, Figure 1.1, p. 4), both marine and nonmarine remains are reworked upward from the Bladen Formation into a lag deposit at the base of the overlying Donoho Creek Formation. Likewise, most of the Cretaceous fossils from Stokes Quarry came from a lag deposit underlying the Pliocene Duplin Formation.

7

SUMMARY AND CONCLUSIONS

The present study reports 275 specimens representing diverse Cretaceous reptiles and dinosaurs from fourteen localities in six counties in eastern South Carolina (Table 1.1, p. 6). These sites range in age from the early middle Campanian (Coachman Formation equivalent, one site) to late middle Campanian (Bladen Formation, one site), early late Campanian (Donoho Creek Formation, three sites and one probable assignment), early Maastrichtian (Peedee Formation, five sites), to the late Maastrichtian (Steel Creek Formation equivalent, one site). The age of the Cretaceous sediments at three of the sites (12, 13, 14) could not be determined because of insufficient stratigraphic data. Because of the fragmentary nature of the material, identification to a specific level was attempted in just nine cases, that is, the turtles *Euclastes wielandi* (Pancheloniidae), *Osteopygis emarginatus* (Macrobaenidae), "*Trionyx*" *halophilus* and "*T.*" *priscus* (Trionychidae), *Peritresius ornatus* (Toxochelyidae), and a new taxon, *Corsochelys bentleyi* (Dermochelyidae); the giant eusuchian crocodile *Deinosuchus rugosus*; and the theropod dinosaurs *Saurornitholestes langstoni* and *Appalachiosaurus montgomeriensis*.

Table 1.1 demonstrates that Stokes Quarry (Site 1), a now-inactive sand mine in Darlington County, was by far the most productive locality, yielding 164 specimens, 60 percent of the material included in this study. Specimens from Stokes Quarry were recovered from spoil piles resulting from commercial excavation; therefore, knowledge of the detailed environmental setting and age of those specimens is lacking. As previously noted, that site also furnished two multituberculate specimens, a partial tooth (ChM PV8700) and a proximal left femur (ChM PV8582), the first Cretaceous mammal remains from South Carolina. Excluding the seven coprolites from the quarry, 164 dinosaurian and reptilian bones and teeth from this site were used in the present study. Although the other thirteen localities reported in this volume are also of considerable importance, Stokes Quarry has provided the largest sample of the Late Cretaceous reptile and dinosaur faunas of South Carolina.

Table 1.1 also shows that turtle bones (66) were the most abundant tetrapod remains from the mid-Campanian deposits at Stokes Quarry, comprising approximately 44 percent of the total. Crocodilian (53) and dinosaur (24) specimens were the second (32 percent) and third (15 percent) most numerous tetrapod vertebrate remains that we report from

Family	Taxon	New Jersey	Delaware Maryland	North Carolina	South Carolina	Georgia	Alabama
Taphrosphidae	*Taphrosphys sulcatus*	M					
	Bothremys cooki	M					
	Bothremys sp.			M-UC	M-UC		M-UC
	Chedighaii sp.			M-UC			
	Chedighaii barberi	M-UC				M-UC	US-LC
Macrobaenidae	*Osteopygis emarginatus*	M	M		M-UC		
Adocidae	*Adocus* spp.	MC-M			M-UC		
	Agomphus pectoralis	M					
Trionychidae	*"Trionyx" halophilus*	MC-M?	M	M-UC	M-UC		
	"Trionyx" priscus	MC-M?			MC-M		
Protostegidae	*Chelosphargis advena*				US-LC		
	Protostega sp.	M-UC					M-UC
	Protostega gigas						US-LC

Figure 7.1 Upper Cretaceous turtles recorded in the eastern and southeastern United States from New Jersey to Alabama. The most diverse faunas so far documented are from Alabama, New Jersey, and South Carolina. Alabama data comes from Zangerl (1953, 1960), Nichols (1988), Hooks (1998), and Kiernan (2002); Georgia data from Baird (1964) and Schwimmer (1986); South Carolina data from this volume; North Carolina data from Miller (1967) and Gaffney et al. (2009); Maryland and Delaware data from Baird (1964, 1986) and Baird and Galton (1981); New Jersey data from Hay (1908), Baird (1964, 1984), Gaffney (1975b), Olson and Parris (1987), Grandstaff et al. (1992), and Gaffney et al. (2006). US = upper Santonian, LC = lower Campanian, MC = middle Campanian, UC = upper Campanian, M–UC = middle and upper Campanian, M = Maastrichtian.

this locality. Stokes Quarry is particularly remarkable for the number of dinosaur specimens that it has yielded, among the larger assemblages of such material known from a single locality on the eastern Late Cretaceous outcrop of the United States.

Published Late Cretaceous turtle remains from the Atlantic and Gulf Coastal Plains have come from seven states (Figure 7.1). The majority of previously known material has come from New Jersey and Alabama; however, with the results published here, South Carolina becomes the state with the third most diverse Cretaceous turtle fauna. In almost all cases, the turtle material was found in predominantly marine strata, and, consequently, marine and euryhaline taxa are most abundant.

As previously noted, the Cretaceous sediments at Stokes Quarry are age-equivalent to the middle Campanian Coachman Formation. As such,

Family	Species						
Dermochelyidae	Corsochelys new species	M-UC			M-UC		
	Corsochelys haliniches						US-LC
Toxochelyidae	Thinochelys lepisosstea						US-LC
	Peritresius ornatus	M	M		M		
	Peritresius sp.					M	M-UC
	Toxochelys sp.				M-UC		
	Toxochelys moorevillensis						US-UC
	Toxochelys barberi						US-UC
	Prionochelys nauta	M					M-UC
	Prionochelys mutatina						US-LC
	Ctenochelys acris						US-LC
	Ctenochelys tenuitesta						US-LC
	Lophochelys venatrix						US-LC
Pancheloniidae	Euclastes wielandi	MC-M	M		M-UC		

they are also of the same age as the Cretaceous deposits at Phoebus Landing on the Cape Fear River in Bladen County, North Carolina. Self-Trail et al. (2004) reported that "The sediments at Phoebus Landing are part of the Tar Heel Formation of North Carolina. They are specifically age-equivalent to the marine Coachman Formation of South Carolina, based on marine calcareous nannofossil and nonmarine palynomorph assemblages." Miller (1967) reported vertebrate remains from Phoebus Landing that included *Squalicorax*, *Trionyx*, *Tylosaurus*, *Leidyosuchus?* (=*Borealosuchus*), and *Ornithomimus?*, taxa that also occur at Stokes Quarry.

Among the taxa reported here, at least four are known to have survived well into the Paleocene of South Carolina. Specimens of *Adocus*, *Osteopygis emarginatus* (Hutchison and Weems 1998), *Leidyosuchus* (= *Borealosuchus*) and *Bottosaurus* (Erickson 1998) were recorded from the Chicora Member of the Williamsburg Formation (Black Mingo Group) near St. Stephen in Berkeley County (Sanders 1998) and are represented by specimens in The Charleston Museum. Consequently, those reports and those included here have now documented the first known continuum of reptilian taxa across the Cretaceous/Paleocene boundary in South Carolina.

REFERENCES

Agassiz, L. 1849. "Remarks on Crocodiles of the Green Sand of New Jersey and on the *Atlantochelys*." *Proceedings of the Academy of Natural Sciences of Philadelphia* 4: 169.

Allmon, W. D., and J. L. Knight. 1993. "Paleontological Significance of a Turritelline Gastropod-Dominated Assemblage in the Cretaceous of South Carolina." *Journal of Paleontology* 67 (3): 355–360.

Baird, D. 1964. "A Fossil Sea-Turtle from New Jersey." *New Jersey State Museum Investigations* 11: 3–26.

_____. 1984. "Evidence of Giant Protostegid Sea-Turtles in the Cretaceous of New Jersey." *The Mosasaur* 2: 135–140.

_____. 1986. "Upper Cretaceous Reptiles from the Severn Formation of Maryland." *The Mosasaur* 3: 63–83.

_____, and G. R. Case. 1966. "Rare Marine Reptiles from the Cretaceous of New Jersey." *Journal of Paleontology* 40 (5): 1211–1215.

_____, and J. R. Horner 1979. "Cretaceous Dinosaurs of North Carolina." *Brimleyana* 2: 1–28.

_____, and P. M. Galton. 1981. "Pterosaur Bones from the Upper Cretaceous of Delaware." *Journal of Vertebrate Paleontology* 1: 67–71.

Barsbold, R. 1976. "On a New Late Cretaceous Family of Small Theropods (Oviraptoridae fam. nov. in Mongolia)." *Doklady Akademia Nauk SSSR.* 226 (3): 685–688 (in Russian).

_____, and H. Osmólska. 1990. "Ornithomimosauria." In *The Dinosauria*, edited by D. B. Weishampel, P. Dodson, and H. Osmólska, 225–244. Berkeley: University of California Press.

Batsch, A. J. G. C. 1788. *Versuch einer Anleitung zur Kenntniss und Geschichte der Their und Mineralien.* Jena, Germany: Akademische Buchhandlung.

Baur, G. 1891. "Notes on Some Little Known American Fossil Tortoises." *Proceedings of the Academy of Natural Sciences of Philadelphia* 43: 411–430.

_____. 1895. "Über die Morphologie des Unterkiefers der Reptilien." *Anatomischer Anzeiger* 11: 410–415.

Bell, G. D. 1997. "Mosasaurs." In *Ancient Marine Reptiles*, edited by J. M. Callaway and E. M. Nicholls, 293–332. San Diego: Academic Press.

Benson, P. H. 1969. "Evidence Against a Large-Scale Disconformity Between the Upper Cretaeous Black Creek and Peedee Formations in

South Carolina." *South Carolina Geological Survey Geologic Notes* 13 (2): 47–50.

Blainville, H. D., de. 1835. "Description de quelques espèces de reptiles de la Californie, précédée de l'analyse d'un système général d'Erpetologie et Amphibiologie." *Nouvelle Annales du Muséum d'Histoire naturelle* Paris 4: 233–296.

Brinkman, D. B., J. J., Densmore, and W. G. Joyce. 2010. " 'Macrobaenidae' (Testudines: Eucryptodira) from the Late Paleocene (Clarkforkian) of Montana and the Taxonomic Treatment of '*Clemmys*' *backmani*." *Bulletin of the Peabody Museum of Natural History* 51(2): 147–155.

Brochu, C. A. 1997. "A Review of *Leidyosuchus* (Crocodyliformes, Eusuchia) from the Cretaceous Through Eocene of North America." *Journal of Vertebrate Paleontology* 17: 679–697.

_____. 2004. "A New Gavialoid Crocodylian from the Late Cretaceous of Eastern North America and the Phylogenetic Relationships of Thoracosaurs." *Journal of Vertebrate Paleontology* 24: 610–633.

Bryan, J. R., D. L. Fredrick, D. R. Schwimmer, and W. G. Siesser. 1991. "First Dinosaur Record from Tennessee: A Campanian Hadrosaur." *Journal of Paleontology* 65 (4): 696–697.

Bybell, L. M., K. J. Conlon, L. E. Edwards, N. O. Frederikson, G. S. Gohn, and J. M. Self-Trail. 1998. *Biostratigraphy and Physical Stratigraphy of the Cannon Park Core (CHN-800), Charleston, South Carolina.* U.S.Geological Survey Open File Report 98-246. Reston, VA: U.S Geological Survey.

Carpenter, K. 1983. "*Thoracosaurus neocesariensis* (DeKay, 1842) (Crocodylia: Crocodylidae) from the Late Cretaceous Ripley Formation of Mississippi." *Mississippi Geology* 4 (1): 1–10.

Carr, T. D., T. E. Williamson, and D. R. Schwimmer. 2005. "A New Genus and Species of Tyrannosauroid from the Late Cretaceous (Middle Campanian) Demopolis Formation of Alabama." *Journal of Vertebrate Paleontology* 25 (1): 116–140.

Christopher, R. A., and D. C. Prowell. 2002. "A Palynological Zonation for the Maastrichtian Stage (Upper Cretaceous) of South Carolina." *Cretaceous Research* 23 (2002): 639–669.

_____. 2010. "A Palynological Zonation for the Uppermost Santonian and Campanian Stages (Upper Cretaceous) of South Carolina, USA." *Cretaceous Research* 31 (2010): 101–129.

Cicimurri, D. J. 2007. "A Late Campanian (Cretaceous) Selachian Assemblage from a Classic Locality in Florence County, South Carolina." *Southeastern Geology* 45 (2): 59–72.

Colbert, E. H., and R. T. Bird. 1954. "A Gigantic Crocodile from the Upper Cretaceous Beds of Texas." *American Museum Novitates* 1688: 1–22.

Coombs, W. P., Jr. 1990. "Behaviour Patterns in Dinosaurs." In *The Dinosauria*, edited by D. B. Weishampel, P. Dodson, and H. Osmólska, 32–42. Berkeley: University of California Press.

Cope, E. D. 1864. "On the Limits and Relations of the Raniformes." *Proceedings of the Academy of Natural Sciences of Philadelphia* 16: 181–183.

———. 1867. "On *Euclastes*, a Genus of Extinct Cheloniids." *Proceedings of the Academy of Natural Sciences of Philadelphia* 19: 39–42.

———. 1868. "On the Origin of Genera." *Proceedings of the Academy of Natural Sciences of Philadelphia* 1868: 242–300.

———. 1869a. "The Fossil Reptiles of New Jersey." *American Naturalist* 3: 84–91.

———. 1869b. *Synopsis of the Extinct Batrachia, Reptilia and Aves of North America.* Transactions of the American Philosophical Society, 14. Philadelphia: American Philosophical Society.

———. 1869c. "Synopsis of the Extinct Reptilia Found in the Mesozoic and Tertiary Strata of New Jersey." In *Geology of New Jersey*, edited by G. H. Cook, 733–738. Trenton: New Jersey Geological Survey.

———. 1869d. "Remarks on *Eschrichtius polyporus*, *Hypsibema crassicaudata*, *Hadrosaurus tripos*, and *Polydectes biturgidus*." *Proceedings of the Academy of Natural Sciences of Philadelphia* 21: 192.

———. 1870. "On the Adocidae." *Proceedings of the American Philosophical Society* 11: 384.

———. 1873. "On *Toxochelys latiremis*." *Proceedings of the Academy of Natural Sciences of Philadelphia* 25: 10.

Currie, P. J., J. K. Rigby, Jr., and R. E. Sloan. 1990. "Theropod Teeth from the Judith River Formation of Southern Alberta, Canada." In *Dinosaur Systematics, Approaches and Perspectives*, edited by K. Carpenter and P. J. Currie, 107–125. Cambridge: Cambridge University Press.

Dollo, L. 1889. "Nouvelle note sur les vertébrés fossiles recémment offerts aux Musée de Bruxelles par M. Alfred Lemonniere." *Societe Belge Géologique Procès-Verbaux* 3: 214–215.

Emmons, E. 1856. *Report of the North Carolina Geological Survey*, 219–221. Raleigh, NC: U.S. Geological Survey.

Erickson, B. R. 1976. "Osteology of the Early Eusuchian Crocodile *Leidyosuchus formidabilis*, sp. nov. 2." *Monograph vol. 2: Paleontology*. St. Paul: The Science Museum of Minnesota.

_____. 1998. "Crocodilians of the Black Mingo Group (Paleocene) of the South Carolina Coastal Plain." In *Paleobiology of the Williamsburg Formation (Black Mingo Group; Paleocene) of South Carolina, USA*, edited by A. E. Sanders, 196–214. Transactions of the American Philosophical Society, 88 (4). Philadelphia: American Philosophical Society.

Fastovsky, D. E. 1985. "A Skull of the Cretaceous Chelonioid Turtle *Osteopygis* and the Classification of the Osteopyginae." *New Jersey State Museum Investigation* 3: 1–23.

Fitzinger, L. J. 1843. *Systema Reptilium*—Fasciculus primus: *Amblyglossae*. Koblenz, Germany: Vindobonae.

Gaffney, E. S. 1975a. "A Phylogeny and Classification of the Higher Categories of Turtles." *Bulletin of the American Museum of Natural History* 155: 389–436.

_____. 1975b. "A Revision of the Side-Necked Turtle *Taphrosphys sulcatus* (Leidy) from the Cretaceous of New Jersey." *American Museum Novitates* 2571: 1–24.

_____, H. Tong, and P. A. Meylan. 2006. "Evolution of the Side-Necked Turtles: The Families Bothremydidae, Euraxemydidae, and Araripemydidae." *Bulletin of The American Museum of Natural History* 300: 1–698.

_____, G. E. Hooks III, and V. P. Schneider. 2009. "New Material of North American Side-Necked Turtles (Pleurodira: Bothremydidae)." *American Museum Novitates* 3655: 1–26.

Gallagher, W. B. 1993. "The Cretaceous/Tertiary Mass Extinction Event in the Northern Atlantic Coastal Plain." *The Mosasaur* 5: 75–154.

Gardner, J. D., A. P. Russell, and D. B. Brinkman. 1995. "Systematics and Taxonomy of Soft-Shelled Turtles (Family Trionychidae) from the Judith River Group (Mid-Campanian) of North America." *Canadian Journal of Earth Sciences* 32: 631–643.

Gervais, P. 1853. "Observations relatives aux reptiles fossiles de France." *Academie Scientifique Paris Comptes rendus* 36: 374–377, 470–474.

Gibbes, R. W. 1850. "On *Mosasaurus* and Other Allied Genera in the United States." *Proceedings of the American Association for the Advancement of Science*, 2nd Meeting, 1849, Cambridge, Massachussetts: 77.

_____. 1851. "A Memoir on *Mosasaurus* and the Three Allied New Genera, *Holcodos, Conosaurus*, and *Amophorosteus*." *Smithsonian Contributions to Knowledge* 2 (5): 1–13.

Gilmore, C. W. 1911. "A New Fossil Alligator from the Hell Creek Beds of Montana." *Proceedings of U.S. National Museum* 41 (1860): 297–302.

_____. 1942. "Osteology of *Polyglyphandon*, an Upper Cretaceous Lizard from Utah." *Proceedings of U.S. National Museum* 92 (3148): 229–265.

Goddard, E. N., P. D. Trask, R. K. De Ford, O. N. Rove, J. T., Singewald, Jr., and R. M. Overbeck. 1984. *Rock Color Chart*. Boulder, CO: Geological Society of America.

Gohn, G. S. 1992. *Revised Nomenclature, Definitions, and Correlations for the Cretaceous Formations in USGS—Clubhouse Crossroads #1, Dorchester County, South Carolina*. U.S. Geological Survey Professional Paper 1581. Washington, DC: U.S. Government Printing Office.

_____, B. B. Higgins, C. C. Smith, and J. P. Owens. 1977. "Lithostratigraphy of the Deep Corehole (Clubhouse Crossroads Corehole 1) Near Charleston, South Carolina." In *Studies Related to the Charleston Earthquake of 1886—A Preliminary Report*, 59–70. U.S. Geological Survey Professional Paper 1028–E. Washington, DC: U.S. Government Printing Office.

Gradstein, F., J. Ogg, M. Schmitz, and G. Ogg. 2012. *The Geologic Time Scale 2012*. Waltham, MA: Elsevier.

Grandstaff, B. S., D. C. Parris, , R. K. Denton, Jr., and W. B. Gallagher. 1992. "*Alphadon* (Marsupialia) and Multituberculata (Allotheria) in the Cretaceous of Eastern North America." *Journal of Vertebrate Paleontology* 12: 217–222.

Gray, J. E. 1825. "A Synopsis of the Genera of Reptiles and Amphibia, with a Description of Some New Species. *Annals of Philosophy* 10: 193–217.

_____. 1827. "A Synopsis of the Genera of Saurian Reptiles, in Which Some New Genera Are Indicated, and Some Others Reviewed by Actual Examination." *Philadelphia Magazine* 2: 54–58.

Haq, B. U., J. Hardenbol, and P. R. Vail. 1987. "Chronology of Fluctuating Sea-Levels Since The Triassic." *Science* 235: 1156–1166.

Harrell, S. D., and D. R. Schwimmer. 2010. "Coprolites of *Deinosuchus* and Other Crocodylians from the Upper Cretaceous of Western Georgia, USA." In *Crocodyle Tracks and Traces*, edited by J. S. Milàn, S. G. Lucas, M. G. Lockley, and J. A. Spielman, 209–213. Albequerque, NM: New Mexico Museum of Natural History Bulletin 51.

Hay, O. P. 1908. *The Fossil Turtles of North America*. Washington, DC: Carnegie Institution of Washington Publication 75.

_____. 1930. *Second Bibliography and Catalogue of the Fossil Vertebrata of North America*. Washington, DC: Carnegie Institute of Washington Publication 390.

Heron, S. D. 1958. "History of the Terminology and Correlations of the Basal Cretaceous Formations of the Carolinas." *South Carolina Division of Geology Bulletin* 2: 77–88.

Hirayama, R. 1992. "Humeral Morphology of Chelonioid Sea-Turtles: Its Functional Analysis and Phylogenetic Implications." *Bulletin of the Hobetsu Museum* 8: 17–57.

_____. 1994. "Phylogenetic Systematics of Chelonioid Sea Turtles." *The Island Arc* 3: 270–284.

_____, and H. Tong. 2003. "*Osteopygis* (Testudines: Cheloniidae) from the Lower Tertiary of the Ouled Abdoun Phosphate Basin, Morocco." *Palaeontology* 46: 845–856.

Hoffstetter, R., and J.-P. Gasc. 1969. "Vertebrae and Ribs of Modern Reptiles." In *Biology of the Reptilia, Morphology A*, edited by C. Gans, A da Bellairs, and T. S. Parsons, 201–310. London: Academic Press.

Holland, W. J. 1909. "*Deinosuchus hatcheri*, a New Genus and Species of Crocodile from the Judith River Beds of Montana." *Annals of the Carnegie Museum* 6: 281–294.

Holtz, T. R., Jr. 1996. "Phylogenetic Taxonomy of the Coelurosauria (Dinosauria: Theropoda)." *Journal of Paleontology* 70: 536–538.

_____. 2004. "Tyrannosauroidea." In *The Dinosauria* (2nd ed.), edited by D. B. Weishampel, P. Dodson, and H. Osmólska, 111–136. Berkeley: University of California Press.

Hooks, G. E., III, 1998. "Systematic Revision of the Protostegidae, with a Redescription of *Calcarichelys gemma* Zangerl, 1953." *Journal of Vertebrate Paleontology* 18: 85–98.

Horner, J. R. 1992. "Cranial Morphology of *Prosaurolophus* (Ornithischia: Hadrosauridae) with Descriptions of Two New Hadrosaurid Species and an Evaluation of Hadrosaurid Phylogenetic Relationships." *Museum of the Rockies Occasional Paper* 2: 1–119.

Huene, F. von. 1914. "Das naturaliche System der Saurischia." *Zentralblatte Mineralogie, Geologie, und Palaeontologie, B* 1914: 154–158.

Hummel, K. 1929. "Die fossilen Weichschildkröten (Trionychia). Eine morphologisch-systematische und stammesgeschichtliche studie." *Geologische und Palaeontologische Abhandlungen* 16: 359–487.

Hunt, A. P., and S. G. Lucas. 2010. "Crocodylian Coprolites and the Identification of the Producers of Coprolites." In *Crocodyle Tracks and Traces*, edited by J. Milàn, S. G. Lucas, M. G. Lockley, and J. A. Spielmann, 219–226. Albequerque, NM: New Mexico Museum of Natural History Bulletin, 51.

Hutchison, J. H., and R. E. Weems. 1998. "Paleocene Turtle Remains from South Carolina." In *Paleobiology of the Williamsburg Formation (Black Mingo Group: Paleocene) of South Carolina, USA*, edited by A.E. Sanders, 165–195. Transactions of the American Philosophical Society, vol. 88, no. 4. Philadelphia: American Philosophical Society.

Huxley, T. H. 1875. "On *Stagonolepis robertsoni* and the Evolution of the Crocodilia." *Quarterly Journal of the Geological Society of London* 31: 423–428.

Iordansky, N. H. 1973. "The Skull of the Crocodilia." In *Biology of the Reptilia, Morphology D*, edited by C. Gans and T. S. Parsons, 201–262. London: Academic Press.

Joyce, W. G., J. F. Parham, and J. A. Gauthier, 2004. "Developing a Protocol for the Conversion of Rank-Based Taxon Names to Phylogenetically Defined Clade Names, as Exemplified by Turtles." *Journal of Paleontology* 78(5): 989–1013.

Kiernan, C.R. 2002. "Stratigraphic Distribution and Habitat of Segregation of Mosasaurs in the Upper Cretaceous of Western and Central Alabama, with an Historical Review of Alabama Mosasaur Discoveries." *Journal of Vertebrate Paleontology* 22: 91–103.

_____, and D. R. Schwimmer. 2004. "First Record of a Velociraptorine Theropod (Tetanurae, Dromaeosauridae) from the Eastern Gulf Coast United States." *The Mosasaur* 7: 89–93.

Konishi, T., and M. W. Caldwell. 2007. "New Specimens of *Platecarpus planifrons* (Cope, 1874) (Squamata: Mosasauridae) and a Revised

Taxonomy of the Genus." *Journal of Vertebrate Paleontology* 27(1): 59–72.

Langston, W., Jr. 1960. "The Vertebrate Fauna of the Selma Formation in Alabama, Part VI. The Dinosaurs." *Fieldiana Geological Memoirs* 3: 313–363.

Lawrence, D. R., and J. P. Hall. 1987. "The Upper Cretaceous Peedee–Black Creek Formational Contact at Burches Ferry, Florence County, South Carolina." *South Carolina Geology* 31 (2): 59–66.

Leidy, J. 1851. "Descriptions of Fossils from the Green-Sand of New Jersey." *Proceedings of the Academy of Natural Sciences of Philadelphia* 5: 329–331.

_____. 1852. "Descriptions of *Delphinus conradi* and *Thoracosaurus grandis.*" *Proceedings of the Academy of Natural Sciences of Philadelphia* 6: 35.

_____. 1856. "Notices of Remains of Extinct Turtles of New Jersey, Collected by Professor Cook, of the State Geological Survey." *Proceedings of the Academy of Natural Sciences of Philadelphia* 8: 303–304.

_____. 1858. "*Hadrosaurus foulkii*, a New Saurian from the Cretaceous of New Jersey." *Proceedings of the Academy of Natural Sciences of Philadelphia* 10: 215–218.

_____. 1865. "Memoir on the Extinct Reptiles of the Cretaceous Formations of the United States." *Smithsonian Contributions to Knowledge* 15(article 6): 1–135.

_____. 1868. "Remarks on *Conosaurus* of Gibbes." *Proceedings of the Academy of Natural Sciences of Philadelphia* 1868: 200–202.

Linneaus, C. 1758. *Systema Naturae* (10th ed., vol. 1). Stockholm: 1–824.

Longrich, N. R., and P. J. Currie. 2009. "A Microraptorine (Dinosauria–Dromaeosauridae) from the Late Cretaceous of North America." *Proceedings of the National Academy of Sciences of the United States* 106 (13): 5002–5007.

Lull, R. S., and N. Wright. 1942. "Hadrosaurian Dinosaurs of North America." *Geological Society of America Special Papers* 40: 1–242.

Lyell, C. 1845. "On the White Limestone of South Carolina and Georgia, and the Eocene Strata of Other Parts of the U.S., with Appendix on the Corals by Mr. Lonsdale." *Proceedings of the Geological Society of London* 4 (1843–45): 563–576.

Lynch, S. C., and J. F. Parham. 2003. "The First Report of Hard-Shelled Sea Turtles (Cheloniidae *sensu lato*) from the Miocene of California, Including a New Species (*Euclastes hutchisoni*) with Unusually Plesiomorphic Characters." *PaleoBios* 23 (3): 21–35.

Makovicky, P. S., K. Yoshitsugu, and P. J. Currie. 2004. "Ornithomimosauria." In *The Dinosauria* (2nd ed.), edited by D. B.Weishampel, P. Dodson, and H. Osmólska, 137–150. Berkeley: University of California Press.

Marsh, O. C. 1872. "Notes on *Rhinosaurus*." *American Journal of Science, Series 34* (20): 147.

_____. 1881. "Principal Characters of American Jurassic Dinosaurs, Part V." *American Journal of Science, Series 3* (21): 417–423.

Matthew, W. D., and B. Brown. 1922. "The Family Deinodontidae, with Notice of a New Genus from the Cretaceous of Alberta." *Bulletin of the American Museum of Natural History* 46: 367–385.

Meylan, P. A. 1987. "The Phylogenetic Relationships of Soft-Shelled Turtles (Family Trionychidae)." *Bulletin of American Museum of Natural History* 186: 1–101.

Miller, H.W. 1967. "Cretaceous Vertebrates from Phoebus Landing, North Carolina." *Proceedings of the Academy of Natural Sciences of Philadelphia* 119: 219–235.

_____. 1968. "Additions to the Upper Cretaceous Vertebrate Fauna of Phoebus Landing, North Carolina." *Journal of the Elisha Mitchell Society* 84 (4): 467–471.

Mills, R. 1825. *Mills' Atlas: Atlas of the State of South Carolina*. Columbia, SC: Robert Mills.

Mitchill, S. L. 1818. "Observations on the Geology of North America; Illustrated by the Description of Various Organic Remains Found in that Part of the World." In *Essay on the Theory of the Earth*, American Edition, edited by G. Cuvier, 319–431. New York: Kirk and Mercein.

Mook, C. C. 1925. "A Revision of the Mesozoic Crocodilia of North America." *Bulletin American Museum of Natural History* 51: 348–377.

_____. 1931. "New Crocodilian Remains from the Hornerstown Marls of New Jersey." *American Museum Novitates* 476: 1–15.

Morton, S. G. 1829. "Geological Observations on the Secondary, Tertiary, and Alluvial Formations of the Atlantic Coast of the United States of

America: Arranged from the Notes of Lardner Vanuxem by S. G. Morton, M.D." *Journal of the Academy of Natural Sciences of Philadelphia* 6 (1): 59–71.

_____. 1834. *Synopsis of the Organic Remains of the Cretaceous Group of the United States*. Philadelphia: Key and Biddle.

Nicholls, E. L. 1988. "New Material of *Toxochelys latiremis* Cope, and a Revision of the Genus *Toxochelys* (Testudines, Chelonioidea)." *Journal of Vertebrate Paleontology* 8 (2): 181–187.

Njau, J. K., and R. J. Blumenschine. 2006. "A Diagnosis of Crocodile Feeding Traces on Larger Mammal Bone, with Fossil Examples from the Plio-Pleistocene Olduvai Basin, Tanzania." *Journal of Human Evolution* 50: 142–162.

Norell, M. A., J. M. Clark, and J. H. Hutchison. 1994. "The Late Cretaceous Alligatorid *Brachychampsa montana* (Crocodylia): New Material and Putative Relationships." *American Museum Novitates* 3116: 1–26.

Olson, S. L., and D. C. Parris. 1987. "The Fossil Birds of New Jersey." *Smithsonian Contributions to Paleobiology* 63: 1–22.

Oppel, M. 1811. *Die Ordnungen, Familien und Gattungen der Reptilien als Prodrom einer Naturgeschichte derselben*, 1–87. Munich: J. Lindauer.

Ostrom, J. H. 1961. "Cranial Morphology of the Hadrosaurian Dinosaurs of North America." *American Museum of Natural History Bulletin* 22 (2): 33–186.

Owen, R. 1842. "Report on British Fossil Reptiles, part II." *Report of the British Association for the Advancement of Science* 11: 60–104.

_____. 1860 (for 1859). "On the Orders of Fossil and Recent Reptiles, and Their Distribution in Time." *Reports of the British Association for the Advancement of Science* 29: 153–166.

Owens, J. P. 1989. "Geologic Map of the Cape Fear Region, Florence 1° x 2° Quadrangle and Northern Half of the Georgetown 1° x 2° Quadrangle, North Carolina and South Carolina." *U.S. Geological Survey Miscellaneous Investigations. Map, I-1948-A, scale 1:250,000*. Washington, DC: U.S. Government Printing Office.

Parham, J. F. 2005. "A Reassessment of the Referral of Sea Turtle Skulls to the Genus *Osteopygis* (Late Cretaceous, New Jersey, USA)." *Journal of Vertebrate Paleontology* 25 (1): 71–77.

Prowell, D. C., R. A. Christopher, K. E.Waters, and S. K. Nix. 2003. "The Chrono- and Lithostratigraphic Significance of the Type Section of the Middendorf Formation, Chesterfield County, South Carolina." *Southeastern Geology* 42 (1): 47–66.

Rivera-Sylva, H. E., E. Frey, and J. R. Guzmán-Guitérrez. 2009. "Evidence of Predation on the Vertebra of a Hadrosaurid Dinosaur from the Upper Cretaceous (Campanian) of Coahuila, Mexico." *Carnets de Geologie/Notebooks of Geology* Brest, Letter 20098/02.

Rogers, R. R., and S. M. Kidwell. 2007. "A Conceptual Framework for the Genesis and Analysis of Vertebrate Skeletal Concentrations." In *Bonebeds, Genesis, Analysis and Paleobiological Significance*, edited by A. A. Rogers, D. A. Eberth, and A. R. Fiorello, 2–64. Chicago: University of Chicago Press.

Romer, A. S. 1956. *Osteology of the Reptiles.* Chicago: University of Chicago Press.

Ruffin, E. 1843. *Report of the Commencement and Progress of the Agricultural Survey of South Carolina for 1843.* Columbia, South Carolina.

Russell, D. A. 1967. "Systematics and Morphology of American Mosasaurs (Reptilia, Sauria)." *Bulletin of the Peabody Museum of Natural History* 23: 1–240.

_____. 1972. "Ostrich Dinosaurs from the Late Cretaceous of Western Canada." *Canadian Journal of Earth Sciences* 9: 375–402.

_____. 1988. "A Checklist of North American Marine Cretaceous Vertebrates Including Fresh Water Fishes." *Occasional Papers of the Tyrrell Museum of Paleontology* 4: 1–58.

Sanders, A. E. 1998. "Paleobiology of the Williamsburg Formation (Paleocene) of South Carolina: Summary and Conclusions." In *Paleobiology of the Williamsburg Formation (Black Mingo Group: Paleocene) of South Carolina, USA*, edited by A. E. Sanders, 255–268. Transactions of the American Philosophical Society, 88 (4), Philadelphia: American Philosophical Society.

_____, and W. D. Anderson, Jr. 1999. *Natural History Investigations in South Carolina from Colonial Times to the Present.* Columbia, SC: University of South Carolina Press.

Sankey, J. T. 2001. "Late Campanian Southern Dinosaurs, Aguja Formation, Big Bend, Texas." *Journal of Paleontology* 75 (1): 208–215.

_____, B. R. Standhardt, and J. A. Schiebout. 2005. "Theropod Teeth from the Upper Cretaceous (Campanian–Maastrichtian), Big Bend, National Park, Texas." In *The Carnivorous Dinosaurs*, edited by K. Carpenter, 127–152. Bloomington: Indiana University Press.

Sawyer, G. T. 1981. "A Study of Crocodilian Coprolites from Wannagan Creek Quarry (Paleocene—North Dakota): Ichnofossils II." *Scientific Publications of the Science Museum of Minnesota* 5 (2): 1–29.

_____. 1998. "Coprolites of the Black Mingo Group (Paleocene) of South Carolina." In *Paleobiology of the Williamsburg Formation (Black Mingo Group: Paleocene) of South Carolina, USA*, edited by A. E. Sanders, 221–228. Transactions of the American Philosophical Society, vol. 88, no. 4. Philadelphia: American Philosophical Society.

Schultze, H.-P. 1995. "Terrestrial Biota in Coastal Marine Deposits: Fossil—Lagerstatten in the Pennsylvanian of Kansas, USA." *Palaeogeography, Palaeoclimatology, Palaeoecology* 119: 255–273.

Schumacher, G.-H. 1973. "The Head Muscles and Hyolaryageal Skeleton of Turtles and Crocodilians." In *Biology of the Reptilia, Morphology D*, edited by C. Gans and T. S. Parsons, 101–199. New York: Academic Press.

Schwimmer, D. R. 1986. "Late Cretaceous Fossils from the Blufftown Formation (Campanian) in Western Georgia." *The Mosasaur* 3: 109–123.

_____. 1997a. "Late Cretaceous Dinosaurs in Eastern USA: A Taphonomic and Biogeographic Model of Occurrences." *Second Dinofest International Proceedings*: 203–211.

_____. 1997b. "Predatory Dominance of Giant Crocodiles on the Late Cretaceous Southeastern Coastal Plain." *Geological Society of America, Abstracts With Programs, Southeastern Section* 29 (3): 68.

_____. 2002. *King of the Crocodylians, the Paleobiology of* Deinosuchus. Bloomington: Indiana University Press.

_____. 2010. "Bite Marks of the Giant Crocodylian *Deinosuchus* on Late Cretaceous (Campanian) Bones." In *Crocodyle Tracks and Traces*, edited by J. Milàn, S. G. Lucas, M. G. Lockley, and J. A. Spielmann,

183–190. New Mexico Museum of Natural History Bulletin, vol. 51. Albequerque, NM: New Mexico Museum of Natural History.

_____, and G. D. Williams. 1993. "A Giant Crocodile from Alabama and Observations on the Paleobiology of Southeastern Crocodylians." *Journal of Vertebrate Paleontology* 13 (Suppl. 3): 56A.

_____, G. D. Williams, J. L. Dobie, and W. G. Seisser. 1993. "Late Cretaceous Dinosaurs from the Blufftown Formation in Western Georgia and Eastern Alabama." *Journal of Paleontology* 67 (2): 288–296.

_____, J. D. Stewart, and G. D. Williams. 1997. "Scavenging by Sharks of the Genus *Squalicorax* in the Late Cretaceous of North America." *PALAIOS* 12: 71–83.

Seeley, H. G. 1887. "On the Classification of the Fossil Animals Commonly Called Dinosauria." *Proceedings of the Royal Society of London* 43: 165–171.

Self-Trail, J. M., R. A. Christopher, and D. C. Prowell. 2002. "Evidence for Large-Scale Reworking of Campanian Sediments into the Upper Maastrichtian Peedee Formation at Burches Ferry, South Carolina." *Southeastern Geology* 41 (3): 145–158.

_____, R. A. Christopher, D. C. Prowell, and R. E. Weems. 2004. "The Age of Dinosaur-Bearing Strata at Phoebus Landing, Cape Fear River, North Carolina." *Geological Society of America Abstracts with Programs, Southeastern Section* 36 (2): 117.

Sissingh, W. 1977. "Biostratigraphy of Cretaceous Calcareous Nannoplankton." *Geologie en Mijnbouw* 56: 37–65.

Sloan, E. 1907. "Geology and Mineral Resources." In *Handbook of South Carolina; Resources, Institutions, and Industries of the State*, 77–145. South Carolina Department of Agriculture, Commerce and Immigration.

_____. 1908. "Catalogue of the Mineral Localities of South Carolina." *South Carolina Geological Survey Series* 4 (2): 1–505.

Sohl, N. F., and J. P. Owens. 1991. "Cretaceous Stratigraphy of the Carolina Coastal Plain." In *The Geology of the Carolinas*, 50th Anniversary Volume, edited by J. W. Horton, Jr. and V. A. Zullo, 191–220. Carolina Geological Society. Chapel Hill: University of North Carolina Press.

Souto, P. R. de F. 2010. "Crocodylomorph Coprolites from the Bauru Basin, Upper Cretaceous, Brazil." In *Crocodyle Tracks and Traces*, edited by J. Milàn, S. G. Lucas, M. G. Lockley, and J. A. Spielman, 201–208. New Mexico Museum of Natural History Bulletin, Albequerque, vol. 51.

Stephenson, L. W. 1907. "Some Facts Relating to the Mesozoic Deposits of the Coastal Plain of North Carolina." In *Notes from the Geological Laboratory, 1906–1907*, 93–99. Johns Hopkins University Circular No. 7. Baltimore, MD: Johns Hopkins University Press.

———. 1912. "Eastern Gulf Region, Georgia, and South Carolina." In *Index to the Stratigraphy of North America*, edited by Bailey Willis, 654–660. U.S. Geological Survey Professional Paper 71. U.S. Geological Survey. Washington, DC: U.S. Government Printing Office.

———. 1923. "The Cretaceous Formations of North Carolina." *North Carolina Geological and Economic Survey* 5: 1–604.

Stewart, J. D. 1978. "Enterospirae (Fossil Intestines) from the Upper Cretaceous Niobrara Formation of Western Kansas." In *Fossil Fish Studies*, 9–16. The University of Kansas Paleontological Contributions, Lawrence, Paper 89. Lawrence, KS: University of Kansas.

Stringer, G. L., and L. King. 2012. "Late Eocene Shark Coprolites from the Yazoo Clay in Northeastern Louisiana." In *Vertebrate Coprolites*, edited by A. P. Hunt, J. Milan, S.G. Lucas, and J. A. Speilmann, 275–309. Albequerque, NM: New Mexico Museum of Natural History Bulletin, 57.

Sues, H.-D. 1978. "A New Small Theropod Dinosaur from the Judith River Formation (Campanian) of Alberta, Canada." *Zoological Journal of the Linnean Society* 62: 381–400.

Sukhanov, V. B. 1964. "Subclass Testudinata, Testudinates." In *Fundamentals of Palaeontology. Amphibians, Reptiles and Birds*, edited by J. A. Orlov, 354–438. Moscow: Nauka.

Troxell, E. L. 1925. "*Thoracosaurus*, a Cretaceous Crocodile." *American Journal of Science* 5 (10): 219–33.

Tuomey, M. 1848. *Report on the Geology of South Carolina*. Columbia, SC: A. S. Johnson.

U.S. Geological Survey. 1909. "Geological Names Committee Meeting Minutes," 802. Washington, DC: U.S. Government Printing Office.

Van Nieuwenhuise, D. S., and W. H. Kanes. 1976. "Lithology and Ostracode Assemblages of the Peedee Formation at Burches Ferry, South Carolina. Stratigraphy of the Carolina Cretaceous." *Southeastern Geology* 10 (4): 201–245.

Vanuxem, L.1826. "Report on a Mineralogical and Geological Examination of the State of South Carolina." In *Statistics of South Carolina, Including a View of its Natural, Civil, and Military History, General and Particular*, edited by Robert Mills, 25–30. Charleston: SC: Hurlbut and Lloyd.

Waldman, M., and W. S. Hopkins, Jr. 1970. "Coprolites from the Upper Cretaceous of Alberta, Canada, with a Description of Their Microflora." *Canadian Journal of Earth Science* 7: 1295–303.

Walker, A. D. 1964. "Triassic Reptiles from the Elgin Area: *Ornithosuchus* and the Origin of Carnosaurs." *Philosophical Transactions of the Royal Society B* 248: 53–134.

Weems, R. E., and L. M. Bybell. 1998. "Geology of the Black Mingo Group (Paleocene) in the Kingstree and St. Stephens Areas of South Carolina." In *Paleobiology of the Williamsburg Formation (Black Mingo Group: Paleocene) of South Carolina, USA*, edited by A. E. Sanders, 9–27. Transactions of the American Philosophical Society, 88 (4). Philadelphia: American Philosophical Society.

_____, W. B. Harris, A. E. Sanders, and L. E. Edwards. 2006. "Correlation of Oligocene Sea Level Cycles Between Western Europe and the Southeastern United States." In *Sea Level Changes: Records, Processes and Modeling—SEALAIX' 06*, edited by G. Camoin, A. Droxler, C. Fulthorpe, and K. Miller, 205–206. Abstract, Book Publication 55. Paris: Association des Sedimentologistes Français.

Weishampel, D. B., and L.Young. 1996. *Dinosaurs of the East Coast*. Baltimore, MD: The Johns Hopkins University Press.

Welles, S. P. 1943. "Elasmosaurid Plesiosaurs with Description of New Material from California and Colorado." *Memoirs of the University of California* 13 (3): 125–254.

_____. 1962. "A New Species of Elasmosaur from the Aptian of Columbia and a Review of the Cretaceous Plesiosaurs." *University of California Publications in Geological Sciences* 44 (1): 1–96.

Whetstone, K., and P. Whybrow. 1983. "A "Cursorial" Crocodilian from the Triassic of Lesotho (Basutoland), Southern Africa." *Occasional Publications of the Museum of Natural History of the University of Kansas* 106: 1–37.

Williams, M. E. 1972. "The Origins of "Spiral Coprolites." *The University of Kansas Paleontological Contributions Paper* 59: 1–19.

Williston, S. W. 1897. "Range and Distribution of the Mosasaurs, with Remarks on Synonymy." *Kansas University Quarterly* 6: 177–185.

Wolfe, J. 1976. *Stratigraphic Distribution of Some Pollen Types from the Campanian and Lower Maastrichtian Rocks (Upper Cretaceous) of the Middle Atlantic States*, 1–18. U.S. Geological Survey Professional Paper 977. Washington, DC: U.S. Government Printing Office.

Zangerl, R. 1948. "The Vertebrate Fauna of the Selma Formation of Alabama. Part I. Introduction, Part II. The Pleurodiran Turtles." *Fieldiana, Geology Memoirs* 1 and 2: 56–277.

_____. 1953. "The Vertebrate Fauna of the Selma Formation of Alabama. Part III. The Turtles of the Family Protostegidae. Part IV. The Turtles of the Family Toxochellyidae." *Fieldiana, Geology Memoirs* 3: 60–277.

_____. 1960. "The Vertebrate Fauna of the Selma Formation of Alabama. Part V. An Advanced Cheloniid Sea Turtle." *Fieldiana, Geology Memoirs* 3: 285–312.

_____. 1971. "Two Toxochelyid Sea Turtles from the Landenian Sands of Erqulinnes (Hainaut) of Belgium." *Institut Royal des Sciences Naturelles de Belgique* 160: 1–32.

Zelenitsky, D. K., F. Therrien, G. M. Erickson, C. L. DeBuhr, Y. Kobayashi, D. A Eberth, and F. Hadfield. 2012. "Feathered Non-avian Dinosaurs from North America Provide Insight into Wing Origins." *Science* 338: 510–514.

INDEX

— Index —

— Index —

S

Santee River, 9–11
Saurischia, 93
Sauropterygia, 69
Saurornitholestes, 96, 127
  Saurornitholestes langstoni, 96, 127
Saurornitholestinae, 96
Science Museum of Minnesota, St. Paul, Minnesota, 45
South Carolina State Museum, Columbia, South Carolina, 45
Sparganeaepollinites uniformis, 28
Squalicorax, 115
  Squalicorax kaupi, 117, 122
Squamata, 72
Stokes Quarry, 3–4, 6, 13, 16, 21–25, 45, 47, 49, 51, 53, 56, 61, 63, 66–67, 69, 72, 74, 80–81, 83, 85, 88, 93, 95–96, 98, 101, 103, 106–107, 111, 114–115, 121–123, 127–129
  age, 21–22
  description, 22–25
  formation, 21–22
  history, 22–25
  Holkopollenites forix, 22
Synopsis of the Organic Remains of the Cretaceous Group of the United States, 10
Systematic Paleontology, 43–107

T

taphonomy, 119–123
Taphrosphidae, 128
  Taphrosphys sulcatus, 128
teiid lizards, 6
Teiidae, 72
Tennessee, 10
Testudines, 6, 45

Theropoda, 93-103
Thinochelys lepisosstea, 129
Toxochelyidae, 63,127, 129
Toxochelys, 63–64, 129
  Toxochelys barberi, 129
  Toxochelys moorevillensis, 63, 129
trace fossils, 109–118
Trionychia, 51
trionychids, 5, 53–58, 127–128
Trionychidae, 53, 127–128
"*Trionyx*" *halophilus*, 53, 55, 128
"*Trionyx*" *priscus*, 56–57, 128
Turbeville, 4, 6, 16, 28–29, 69, 80, 87–88, 98
  age, 28–29
  Clarendon County, 3–4, 6, 16, 28–29, 69, 80, 87–88, 98
  description, 28–29
  formation, 28–29
  history, 28–29
Tylosaurus, 5, 74–75
type locality, 34–35, 58
Tyrannosauroidea, 98

U

United States Geological Survey, 9
United States Topographical Engineers, 14
Upper Cretaceous collections, history, 1–6
USGS. *See* United States Geological Survey

V

Vanuxem, Lardner, 9
Varanidae, mosasaurs, 6

W

Waccamaw River, Horry County, 6, 39–40, 80

156

www.ingramcontent.com/pod-product-compliance
Lightning Source LLC
Chambersburg PA
CBHW061753260326
41914CB00006B/1094